P9-EDA-094

H.J. SAVORY
Senior Lecturer in Geography
Borough Road College, Isleworth

A Geography of Wales

CONTENTS

DA
731
S38

CAMBRIDGE · AT THE UNIVERSITY PRESS · 1968

Published by the Syndics of the Cambridge University Press
Bentley House, 200 Euston Road, London, N.W.1.
American Branch: 32 East 57th Street, New York, N.Y. 10022

© Cambridge University Press 1968

Library of Congress Catalogue Card Number: 67-16 791
Standard Book Number: 521 06922 X

ACKNOWLEDGEMENTS

The author wishes to acknowledge the help given by the following:
Mr Edward Davies of Maes Mynach, Mr E. G. Phillips, Mr J. Jones, School House, Talgarth, Mr Stephens of Neuadd, Nr. Talgarth, Mr Islwyn Evans, Public Relations Officer of the South Western Division of the National Coal Board, Mr G. Percival, Assistant General Manager of the Industrial Estates Management Corporation for Wales, Mr L. Williams, Mr Goatman and Mr P. Nunan of Richard Thomas & Baldwins Ltd., Mr K. J. Hilton, Director of the Lower Swansea Valley Project, Mr C. A. Risbridger C.B.E., General Manager of the City of Birmingham Water Department, Mr Hugh Llewellyn, Information Officer, City of Cardiff, Mr P. S. P. Fell, Public Relations Manager, Gulf Eastern Coy., Mr H. H. Mew, Public Relations Officer, Regent Oil Coy., the Public Relations Officer of the North-west Region of the Central Electricity Generating Board, Mr H. Sprague, Mr Bert Isaac for preparing Figure 2, and Mr. J. Voller.

Thanks are also due to:
The National Coal Board for permission to reproduce Figs. 20, 22, and 23, the Industrial Estates Management Corporation for Wales for Fig. 26, Richard Thomas & Baldwins Ltd. for Figs. 29 and 30, Aero Films Ltd. for Figs. 1, 9, 45 and 46, J. Allan Cash for Figs. 6 and 10, H. Tempest Ltd. for Figs. 19 and 33, Terence Soames Ltd. for Fig. 35 and Romley Marney for Fig. 17.

Printed in Great Britain by
Hazell Watson & Viney Ltd, Aylesbury, Bucks

1 | FLINT

On the coastal route from Chester into North Wales, Flint is the first of several Welsh castle towns, though its historic features have been overshadowed by modern industry.

Fig. 2 is the record of a walk which started from the edge of the Dee estuary, passed through Flint, followed the Cornist Road and finished on Halkyn Mountain. The first part of the route can be seen on Fig. 1 and most of the remainder, as far as Pentre Halkyn (Map Ref. 203724), on Map A. The record takes the form of a field notebook and starting at the bottom of the page, can be read facing the direction which the route follows. Changes in direction are not given.

This short walk reveals some interesting geographical contrasts. It starts at the margin of the Dee estuary on recently deposited alluvium and continues across a sequence of rocks from Coal Measures through Millstone Grit to the Carboniferous Limestone on Halkyn Mountain (Fig. 3). A similar succession of rocks occurs in South Wales and in a number of other parts of Britain. Included in it are rocks of great economic importance.

1. Air photograph of Flint

8. Halkyn Mountain. Open, stony heath with no surface water. Limestone outcrops. Some sheep grazing. Many hollows and some old shafts and lead mines

Large quarries and lime works. Limestone is used by the steel works at Shotton and lime by Chemical works on Merseyside.

Pentre Halkyn, former mining village is 700' above sea level, on steep side of Halkyn mountain

A55 road carries traffic by-passing coastal towns.

Wooded, steepsided valley with narrow level floor.

Nant y Flint ←

7. Layers of sandstone, part of Middle Coal Measures, outcrop on roadside and dip steeply towards coast.

The road descends steep Eastern side of valley. Wats Dyke, similar to Offa's Dyke, a former boundary between the Welsh and Saxons runs along this side of valley.

6. Regular long fields S.W. of Flint. Deep brown soils with much arable land.

← The Cornist Road is in a deep cutting through clay and gravel containing roughly rounded pebbles and boulders of granite and other rocks not native to the Flint area.

5. Modern housing estates on outskirts of Flint encroaching on fields.

Old London Road

→ Swinhiards' Brook

4. Regular streets of old town.

Football ground on flat land near brook adjoining Deeside rayon works

Main Road A 548

Main Railway, Chester – Holyhead.

Beyond the moat the land rises to a slightly higher level than the saltings.

2. Flint Castle. Built by Edward I in 1277, on an outcrop of sandstone. At that time the sea came right up to the wall and provisions were unloaded from ships.

3. Courtaulds Castle rayon mill, an example of the chief industry of Flint. The factory lies outside the old part of the town

1. Saltings – waterlogged alluvium liable to flooding at high tides.

Dee Estuary.

2. **Field notes on the Flint area**

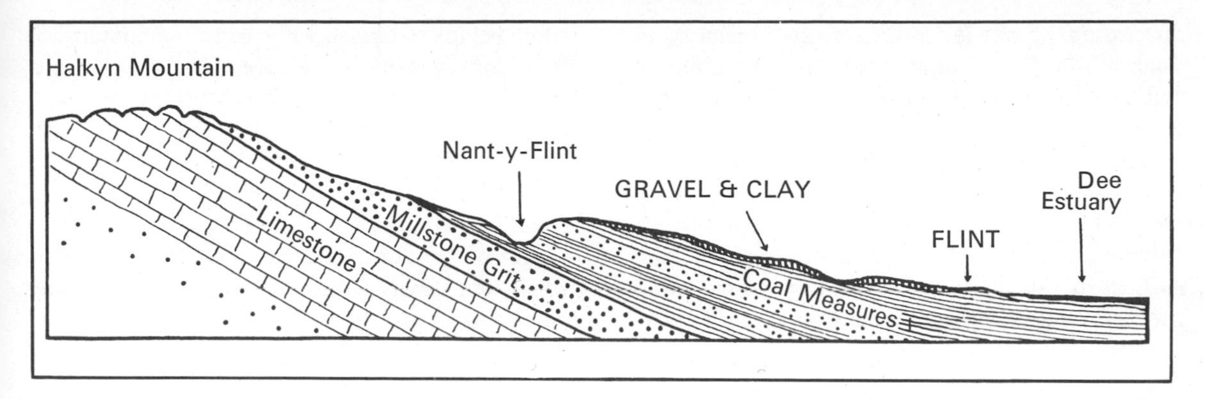

3. Section from Flint to Halkyn Mountain

Along the first half of the route, these older rocks are covered by uneven masses of clays and gravels. The Cornist Road cuts through this hummocky ground and some of the pebbles and boulders found there can only have come from the Lake District or other areas to the north. They were carried into the Flint area by the slowly moving sheet of ice which covered most of Britain during the Ice Age. Such deposits, known as glacial drift, are common in Wales, especially in valleys and lowlands. Usually, however, the drift contains boulders of local rather than distant origin.

The best farm land is on soils formed from the glacial drift, in the area between Nant-y-Flint and the sea. On Halkyn Mountain, above 800 feet, soils are thin and poor and the land is used for rough grazing.

Two valuable minerals occur in the Flint area. Coal is found in seams in the coastal zone and lead in veins through the limestone. Both lead and coal mining have died out, but old mine shafts are numerous. There were lead mines on Halkyn Mountain in Roman times and the ore was worked, on and off, until the most recent mine closed in 1955. The presence of coal was one reason why Flint and neighbouring towns became industrialized. The area forms part of the North Wales coalfield. In the part of the coalfield bordering the Dee estuary in Flintshire very little coal is now mined, but further south in Denbighshire mining is still active. The main industry of Flint today is the manufacture of rayon textiles. The textile factories use water from the local streams and import wood pulp, their chief raw material. Papermaking is another industry in

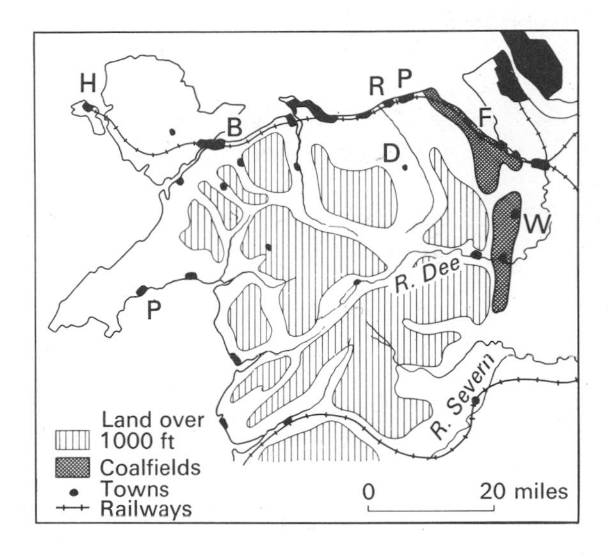

4. North Wales

the area which has similar requirements.

The castle and town of Flint were established by King Edward I as a step towards the conquest of Wales. The castle was at that time on the coast and provisions were supplied by sea. The town, under the protection of the castle, was rectangular in shape, and still has a "gridiron" street pattern. The modern parts of the town have grown outside the old centre and consist mostly of factories on one side and housing estates on the other.

From Flint, the main road and rail routes continue along the coast of North Wales. West of the coalfield (Fig. 4) industrial towns give way to a succession of seaside resorts from Prestatyn to Llandudno. Every summer there is a big influx of holiday-makers mainly from Lancashire and the English Midlands.

The chain of castles built by King Edward I continues along the same route. Conway castle overlooks the Conway estuary at a point where three bridges cross the river into Caernarvonshire—Telford's road bridge, the railway bridge and a modern road bridge. Over the Conway, the landscape begins to change. The rugged mountains of Snowdonia project to the edge of the sea. Welsh is spoken in the towns and countryside. Long ago this area was the stronghold of the north Welsh kingdom of Gwynedd. At its heart were the mountains of Snowdonia, barring the way to Llŷn and Anglesey, where most of the people lived in small scattered hamlets. Caernarvon castle, situated where the Llŷn peninsula begins, was the last major link in the chain.

Exercises Map A should be used in answering questions 1 to 10.

1. Two kinds of minerals have been worked in the area covered by the map. What are they and what evidence of each does the map give?
2. What kinds of industry are shown on the map apart from mining?
3. In which general direction would you be going if travelling (a) along the railway starting at the right-hand side of the map, (b) from Flint, up Cornist Road? Describe the journey along each of these routes.
4. Describe the course and valley of Swinchiard Brook (including Nant-y-Flint) from its upper part to the point where it enters the sea. What is the name for the feature at its mouth and how do you explain it?
5. (a) What use is made of streams in this area? (b) What are the two most common directions followed by streams shown on the map? Suggest a possible explanation.
6. (a) Draw a sketch map of Flint showing the castle, the town and the industries. (b) What were the advantages of the position of Flint when the castle was built, and what are they now?
7. Describe the distribution of woodland. Formerly there was much more woodland. Why do you think there is so little today?
8. Locate on Map A the stages numbered 1 to 7 on Fig. 2.
9. Using the contours on Map A, draw a section from Flint to Halkyn. On your section show the rock formations (with the aid of Fig. 3). Above the section show the following features in their right position: (a) the town of Flint, (b) main roads and railway, (c) woodland, (d) enclosed farm land, (e) mines and quarries. How many of these features are related to rocks and to the relief, as shown by the shape of the section?
10. The kind of diagram you have made in answering Question 9 is called a transect. It shows changes in geography across different kinds of country. If possible, plan a route of a few miles on a map of your own area. Follow it on the ground and record, as on Fig. 2, what you notice about houses, industries, farming, vegetation, soils and relief. If rocks are exposed in quarries or cuttings, describe them. Show the information on a diagram similar to that made in answer to Question 9.
11. (a) Find on the photograph (Fig. 1) the first part of the route recorded on Fig. 2. Find the castle, the textile mills, the main railway and road and the football ground. Note any major features which appear on the photograph but not on Map A. (b) Using Fig. 1, draw a sketch plan of the castle and show on it the former coastline. Part of the area covered by the thirteenth-century castle town appears on the photograph. Find its limits.
12. Using an atlas, identify as many as possible of the towns initialled on Fig. 4. Which of the towns are (a) seaside resorts, (b) industrial towns?

2 | SNOWDON

The sketch shown in Fig. 5 was drawn from the east side of Nant Gwynant looking west towards Snowdon. Some features in this impressive landscape have been picked out and emphasized; many details have been omitted. With the aid of the sketch it is possible to look at Snowdon and Nant Gwynant and to attempt an explanation of what can be seen. Most of the area is included in Map B.

The mountains are rugged, with crags and rock boulders. Most of the rocks are very hard. Slate occurs in many parts of Snowdonia and much of Snowdon itself is made of ancient volcanic lavas and the compressed dust which originally came from volcanic eruptions between 500 and 600 million years ago. The hardness of the rock is one of the reasons why Snowdon attracts climbers.

Steep edges, Grib Goch and Crib-y-ddysol, lead towards the summit and beneath them are deep hollows containing lakes. Beneath the Snowdon summit (Yr Wyddfa) is Glaslyn and at the foot of Grib Goch and Y Lliwedd lies Llyn Llydaw, which is now a reservoir feeding the power station. The hollows are of the type known in this part of Wales as "cwms", and in the Alps as "cirques"; the lakes which occupy them are "cirque" lakes or tarns. In Fig. 6 the two walkers are following a track along the edge of Llyn Llydaw, leading towards an old copper mine which formerly extracted ore from mineral veins running through the mountain. The greenish-blue colour of the lakes is due to their copper content. The sketch (Fig. 5) shows that a third hollow, Cwm Dyli, existing at a lower level, contains only a number of boggy depressions.

A pipe-line from Llyn Llydaw runs across Cwm Dyli, then descends sharply towards the power station, which depends on the very heavy rainfall

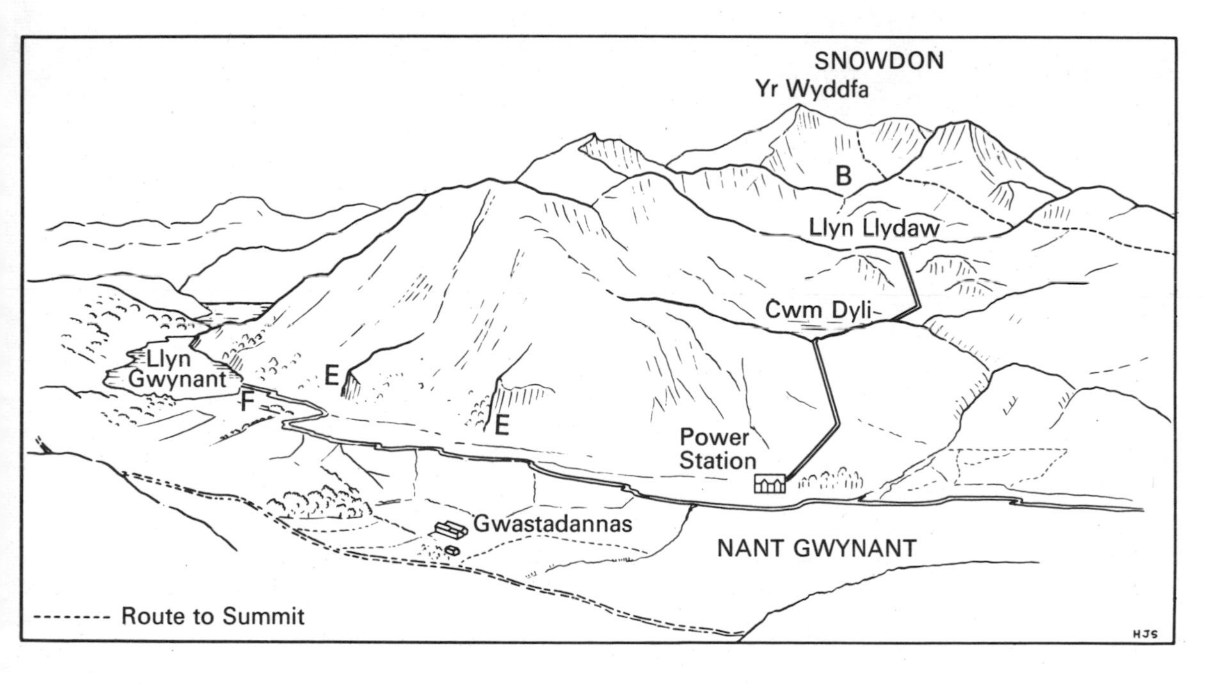

5. Sketch of Snowdon and Nant Gwynant

6. Snowdon and Llyn Llydaw

over Snowdon. The catchment area consists only of the two upper basins of Llyn Llydaw and Cwm Dyli, but their rainfall of over 150 inches a year is sufficient to run the comparatively small power station which is one of the oldest in Wales. Although Wales owes most of its rainfall to depressions which move in from the west, the very heavy rainfall in mountain areas such as Snowdonia is partly caused by the air being forced to rise over the mountains. When this happens the air becomes cooler and the moisture in it condenses to form cloud or rain.

Along the floor of Nant Gwynant, the Afon Glaslyn winds its way towards Llyn Gwynant. Some crags rise steeply from the edge of the valley (Fig. 5, E), but the valley bottom is smoother and less rocky than the mountains. The flatter parts of the valley are formed of sediment laid down by the Afon Glaslyn as it entered the lake. This process is still going on as the alluvial delta shows (Fig. 5, F). Llyn Gwynant itself fills a hollow in the valley floor.

To understand the features shown in Fig. 5, one needs to picture the events of the Ice Age at a period when the hollows and valleys contained glaciers. Fig. 7 is a reconstruction of the landscape at that time which, although imaginary, is based upon evidence to be found on Snowdon today, for example, rocks scratched by ice, boulders carried from one place to another and crags worn smooth on the side from which the ice came. Comparison with a photograph of alpine glaciers today would show many similar features.

Snow from the sides of the mountain at A (Fig. 7) accumulated in the cwm or cirque at B. The thickness was sufficient to compress the lower layers into ice, which poured into the lower cirque at C. The addition of further snowfalls from the side of Y Lliwedd maintained a glacier which flowed via Cwm Dyli into Nant Gwynant to join other ice from the head of the valley near Pen y Pass (see Map B). The action of the ice in the hollows at B and C deepened them and excavated the sides, so creating the cirques which now contain Glaslyn and Llyn Llydaw. The main glacier widened the valley which is now Nant Gwynant (D) and plastered the floor of the valley with drift. In places, E, the glacier removed or "truncated" former spurs jutting into the valley so that steep crags remain. A comparison of Figs. 5 and 7 shows where these crags occur. In some places the glacier hollowed out the floor of the valley and Llyn Gwynant today occupies such a hollow. Since the end of the Ice Age, streams have removed some of the effects of ice and filled, or partially filled, some of the valley lakes with alluvium, but the effects of the ice still show up vividly.

Glaciated mountain country of this kind is difficult to farm. Even where there is soil, the heavy rainfall hinders the growing of any crops except grass and oats. Some centuries ago, farmers used to bring their cattle and sheep into the mountains in summer only and the summer farm was called the *hafod*. The permanent winter farms were near the coast. Note the number of farms in and around Nant Gwynant which have the prefix "hafod" in their names.

Today most of the farms around Snowdon are sheep farms. Gwastadannas (Fig. 5. Map Ref. 656537) has about 1,600 sheep and 2,000 acres of land, most of which is on the mountains. The sheep come down from the mountains in December and as many as possible then live on hay cut during the summer and on the fields in Nant Gwynant. This is not enough to feed the whole flock, so some sheep are sent to farms near the coast in the Llŷn Peninsula. In April the sheep go back on to the mountains. The scattered farms in the valleys support few people.

Some of the towns which lie under the shadow of the mountains, for example Llanberis, Bethesda

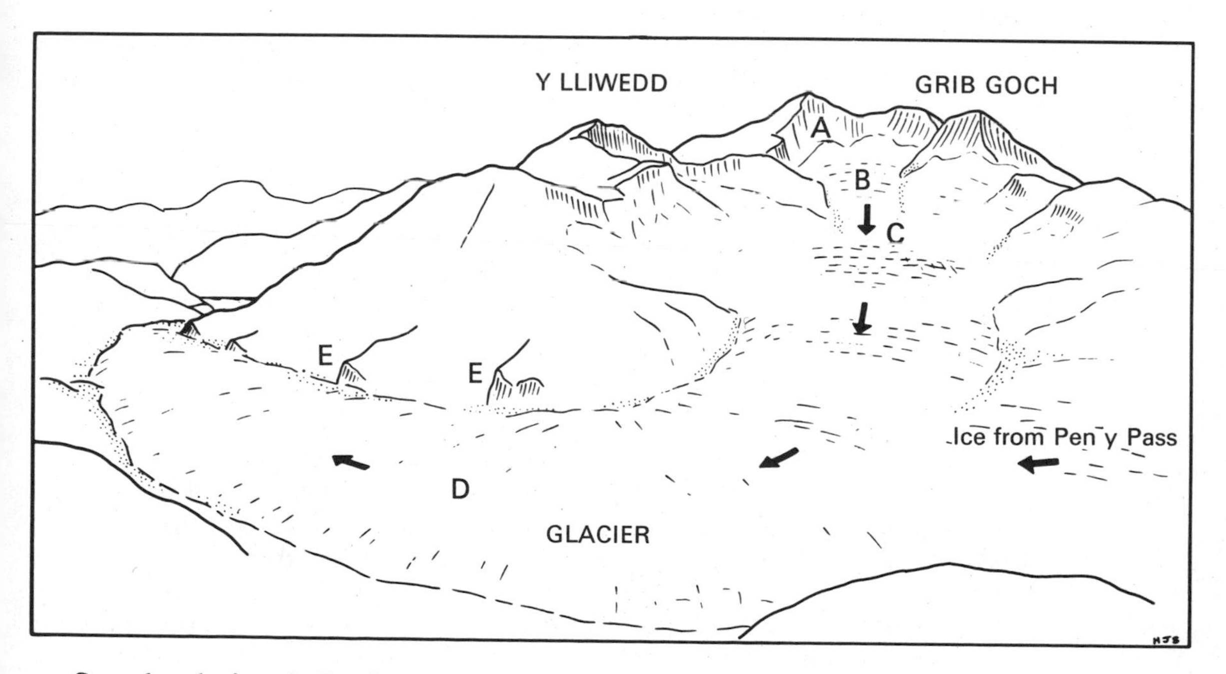

7. Snowdon during the Ice Age

and Blaenau Ffestiniog, began by making their living from slate quarrying. The Dinorwic quarry (Map B), near Llanberis, is one of the largest in the world. The slate is quarried along shelves cutting into the mountain. A great deal of the rock is tipped as waste, while the slate selected is cut into sheets about three inches thick. If required for roofing the slate is sawn into smaller sections and split to widths of about one sixth of an inch. Other roofing materials have now largely replaced slate, and although it is still quarried for building and industrial purposes, it has been for some years a declining industry.

One of the greatest assets of North Wales is its scenery. Snowdonia is a National Park and tourists visit the area in great numbers. The crags attract rock climbers at all seasons and there are many camping sites. Bryn Gwynant, one of the Youth Hostels shown on Map B (Map Ref. 642513), is also a centre for field work, from which parties set out to study on the spot some of the things described in this chapter.

Exercises Exercises 1 to 7 are based on Map B.

1. Describe the route you would follow from the Youth Hostel near Llyn Ogwen (Idwal Cottage), Map Ref. 649604, to the one at Bryn Gwynant. What is the distance? To what height would you climb? Where would be the most difficult part of your walk?
2. (a) Give examples of (i) a cirque, (ii) a deep U-shaped valley;
   (b) name lakes which are (i) cirque lakes or tarns, (ii) ribbon lakes in the valleys, and explain how these two kinds of lakes were formed;
   (c) Llyn Padarn and Llyn Peris were formerly one lake. Explain how they became separated.
3. Name three ways in which people in the Snowdon area earn their living and state the places on the map where they work.
4. State the number of different routes to the summit of Snowdon shown on the map and briefly describe each of them. Which route is shown on Fig. 5?
5. Draw a simple sketch map of the Snowdon area showing (a) ridges leading to the summit, (b) cirques, (c) by means of arrows, the direction in which glaciers moved during the Ice Age.
6. Find on the map the point from which the sketch (Fig. 5) was drawn, and the direction in which the author was looking.
7. Draw a simple section from the milestone (M.S.) at Map Ref. 659540 to Llyn Ffynnon (Map Ref. 591554). With the aid of Fig. 5, show on the section the main physical and human features.
8. Draw your own version of the sketch in Fig. 5. On your sketch, name the following by checking positions on Map B—Snowdon, Y Lliwedd, Grib Goch, Llyn Llydaw, Cwm Dyli power station and the pipe leading to it.
9. What has caused the following features shown in Fig. 5: (a) the cirque in which Glaslyn lies at B, (b) Llyn Gwynant, (c) the crags at E, (d) the plains at F?
10. (a) What kind of farming is carried out in the Snowdon area? (b) Why is it a difficult area to farm? (c) What does the name *hafod* suggest? (d) What activities on a farm in Nant Gwynant today resemble those of the *hafod* in the past?
11. One of the best ways in which to record your observations of a landscape is by sketching. At first the results may not be very good, but in doing it you will look more keenly at the scene and soon improve with practice. When you sketch, draw the main outlines of the mountains, the streams and lakes, the farmhouses and the outlines of the fields and other man-made features like the pipeline to the power station from Cwm Dyli. Do not try to draw too many details. Then, using a map, name the things you have drawn and write on the sketch the direction in which you were looking when you drew it. Write also brief descriptions or explanations of the features you have shown. A collection of such sketches will help you to make up your own account of the geography of an area. Why not begin with the district in which you live?

3 | THE RHEIDOL AND ELAN VALLEYS

A journey up the Rheidol valley from Aberystwyth takes us through a spectacular gorge into the Plynlimon uplands where many changes have been taking place in the use of land and water. Immediately upstream from Aberystwyth the Rheidol river flows swiftly in broad bends or meanders (Fig. 8). Gravel spread over the fields shows that the river sometimes floods the valley floor. Most of the hamlets and farms are above the level of the flood plain.

The photograph (Fig. 9) shows part of the valley at Cwm Rheidol, about eight miles from Aberystwyth. At this point the valley is much more pronounced and the river has been dammed, creating a long winding lake to run a small power station below the dam. In the foreground is a larger power station (Fig. 9, E), part of the Rheidol Power Scheme. A tunnel and pipeline lead into it from the side of the valley. The tunnel brings water from Dinas reservoir about ten miles upstream. On the opposite side of the valley the Rheidol railway, much used by tourists, climbs towards Devil's Bridge. Notice the woods and forest plantations on the sides of the valley where the land is too steep for agriculture.

Upstream from Cwm Rheidol the sides of the valley close in as the river passes through the deep gorge near Devil's Bridge. Along this stretch the river drops 500 feet in two and a half miles. Above the gorge the valley of the Rheidol is broader and several tributaries enter from the Plynlimon range (see Figs. 8 and 10). The streams gradually erode the hills of the Plynlimon range and the river has carried pebbles of grit and shale from the upper basin to the valley below the gorge. In the Rheidol valley the Central Electricity Generating Board has harnessed this source of power through a series of dams and power stations connected by pipe lines (Fig. 8). The dam at Nant-y-Moch (Fig. 8) retains the main body of water which feeds the Dinas reservoir and power station. From Dinas, water flows through a tunnel to the main power station at Cwm Rheidol, by-passing the Rheidol gorge. The pressure at this station is created by the difference in level between the Rheidol above and below the gorge.

Devil's Bridge

Notice in Fig. 8 the gorge and sudden bend in the river at Devil's Bridge. The upper Rheidol, probably not more than a million years ago, flowed southward parallel to the sea coast and joined the River Teifi. After an uplift of the land surface, the lower Rheidol stream cut back from the coast into the hills and "captured" the upper part of the Ystwyth, now the upper Rheidol, at Devil's Bridge. The change in direction of the river at that point is known as the "elbow of capture". The old valley of the Teifi extends as a dry valley for two miles south of Devil's Bridge. The Ystwyth had itself previously captured the upper Teifi.

The Use of the Uplands

Only fifteen miles from the upper Rheidol, on the eastern side of the watershed, are the Elan reservoirs which supply water for over 1,200,000 people in and around Birmingham. An average family consumes nearly fifty gallons of water a day and the manufacture of every car made in Birmingham requires 200,000 gallons. As both population and car production increased, a new dam in the Claerwen valley, next to the Elan valley, was opened in 1962 to meet further demands from the English Midlands. Over the 71 square miles of land which drain into the Elan reservoirs, the average annual rainfall is nearly 70 inches. Moreover, much of this falls in the winter half of the year when there is little loss due to evaporation. From the point of view of the townsman, the water supplies from these uplands are the most valuable of its resources.

The flooding of Nant-y-Moch displaced only one

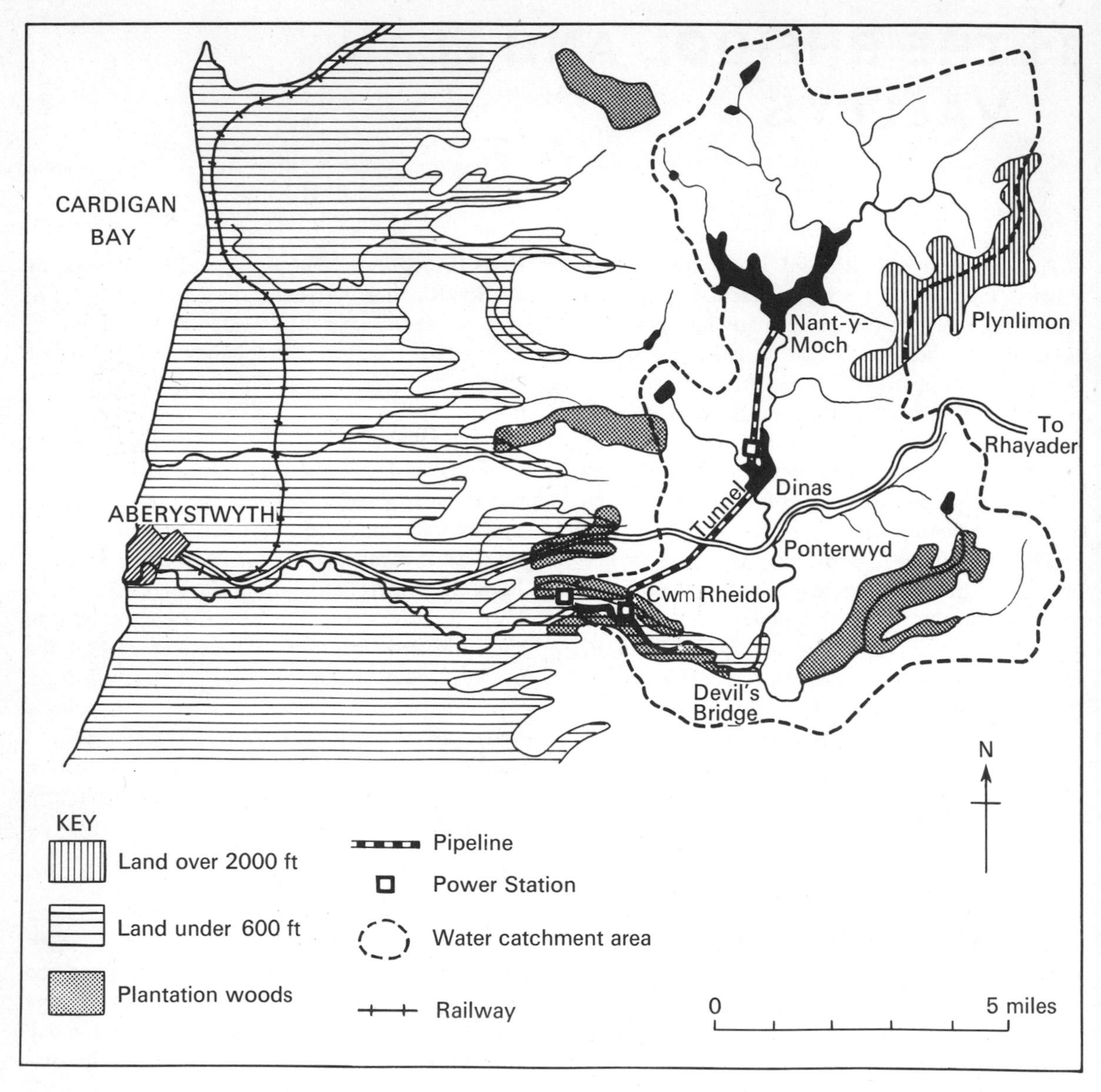

8. The Rheidol Basin

farm run by two brothers both over 70 years of age, and the Electricity Board built a new farm for them. When the first Elan reservoirs were built, between 1894 and 1904, over 20 farms and cottages were moved.

Formerly there were many upland farms; people lived rather isolated lives and farmers travelled on horseback. But during the last hundred years improvements in living conditions and amenities in the towns and villages of the lowlands have led younger generations to leave the farms, many of which have been abandoned. Most of these farms were a long way from main roads and could not be furnished with modern facilities such as electricity and piped water except at considerable cost. To use a car meant rebuilding a road to the farm. The farm shown in Fig. 10 is a typical and traditional upland farm which has survived because it is near a good road. It is near the edge of the moorland and by a stream, but has little shelter. The farm build-

9. Air photograph of Cwm Rheidol

ings are joined together to make a compact long building.

Metal mining, which used to take place in the hills of Cardiganshire, has also died out. For example, the little lead mining village of Ystumtuen, a few miles from the Rheidol, is now almost deserted and the crumbling houses are used for storing hay.

What is the future of these uplands? One solution is the amalgamation of farms to make large sheep ranches. Where this has taken place, the farmers themselves generally do not live in the uplands, but travel there by car. Not all the land is suitable for farming, however. Some can be used for forestry, and the Forestry Commission has planted large areas in Central Wales (see Fig. 8).

Thus, although not many people want to live on these uplands, there is still an active demand for land for three main purposes: sheep ranching, forestry and water supply.

10. **Rheidol Valley near Ponterwyd**

Exercises 1. Using the descriptions and photographs in this chapter, describe briefly three different parts of the Rheidol valley.

2. Study Fig. 9: (a) what use is made of the land at A on the plateau, at B and C on the sides of the valley and at D on the floor of the valley?; (b) what do you notice about the course of the river and the shape of the valley?

3. The following figures show the heights and distances from the sea of certain points in the Rheidol valley:

Place	*Distance from sea*	*Height*
Cwm Rheidol dam	8 miles	150 feet
Elbow of capture (Devil's Bridge)	11 miles	250 feet
Ponterwyd	13½ miles	750 feet
Dinas dam	14 miles	800 feet
Nant-y-Moch dam	17 miles	1,100 feet
Headstream	21 miles	1,750 feet

(a) Draw a simple profile showing the fall in level of the river from its source to the sea and mark on it the positions of the dams and the gorge at Devil's Bridge; (b) explain the shape of the profile.

4. Draw a simple diagram to show how the Rheidol Power Scheme works; add explanatory notes. The capacity of the power stations is as follows:
1. Dinas—13,000 kilowatts, 2. Cwm Rheidol—42,000 kilowatts, 3. Felin Newydd (just below dam shown on Fig. 9)—1,000 kilowatts. (Use information on the height of the dams given in Question 3).

5. An old farmer in the upper Rheidol valley is describing changes which have taken place there in his lifetime. Write down what you imagine he might say.

6. Describe three ways of using the upland areas of Central Wales.

7. Describe and illustrate with sketches the valley and course of a stream or river in your own area. What variations occur in the shape of the valley as you go along it? Has the stream been used for any economic purpose?

8. (a) If the weather is recorded at your school, examine the rain gauge and find out how it works. What is the annual average rainfall in your district? (b) Find the approximate area of your school grounds and calculate how many gallons of water fall on it in an average year. (Assume for this purpose that a gallon of water occupies a space of roughly two square feet and a depth of one inch.) How many people would this supply in a year?

9. Find out what you can about the Elan reservoirs and the pipeline which carries water to the Birmingham area.

4 | A CREAMERY AND DAIRY FARM

At 9 a.m. a driver pulls up his lorry at the entrance to Maes Mynach farm. The farm itself is not within sight, but on a wooden platform are milk cans containing up to ten gallons each.

The driver loads them on to his lorry and goes on to pick up milk from several more farms before he returns to the large modern creamery at Felin Fach in the Vale of Aeron (see Fig. 11). While he has been on this route, other lorries have been bringing in milk from about 1,300 scattered farms in this part of Cardiganshire. Fig. 11 shows the location of the creamery and the many lanes. Other creameries at Newcastle Emlyn, Lampeter and Pont Llanio collect from adjoining areas. At the creamery the milk is manufactured into butter. Surplus milk is delivered by tanker lorry to London and the industrial areas of South Wales. Dried and powdered milk are also produced for the London market and for export when required.

There are other creameries and milk depots in west and South Wales, particularly in the Vale of Towy around Carmarthen. The latter is on the main railway route to London and partly for this reason was one of the first areas in west Wales to supply milk for the London areas. Two developments brought about the spread of dairy farming; the growth of road transport to districts further away from the main railway and the establishment of creameries by the Milk Marketing Board and private firms. The creameries brought into existence local centres demanding milk and giving the farmers a regular and assured source of monthly income. Road transport made it possible to carry the milk daily to those centres. Thus the majority of farmers in areas within range of creameries changed from mixed farming and stock raising to dairy farming. Because of the connection with the London market some Welshmen have emigrated to London to take part in the commercial side of dairying. There are over forty dairymen named Davies and over thirty named Evans listed in the London telephone directory.

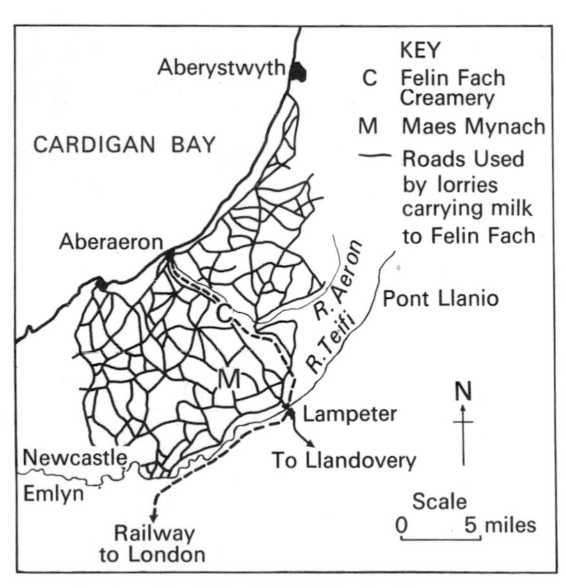

11. Area served by Felin Fach Creamery

Maes Mynach

One of the farms supplying the creamery is Maes Mynach, about six miles south of Felin Fach. A map of the farm, which covers 184 acres, is given in Fig. 12. On a visit to the farm we came down the hill from the small hamlet marked A on Fig. 12, and at the gate where the Milk Marketing Board lorry collects the milk (B), turned back along the lane to Maes Mynach (C). There the farmer, Mr Edward Davies, greeted us and gave us the plan of the farm shown in Fig. 13. Most of the buildings are madc of stone with the exception of the modern barns (marked 7 on Fig. 13).

Opposite the farmhouse the forty black and white Friesian cows are gathered in the collecting yard for the evening milking (see Fig. 13). In the cowshed the modern metal stalls run the length of the building and behind them is a passage from which the cattle can be fed. The cows know their own stalls and go to them. The dairyman cleans and sterilizes the milking machine and cans. After the milk has been drawn from the cows by the milking machine, it is passed into the cooling apparatus. The cooling enables it to stay fresh for much longer. Milking is a routine that goes on twice a day, in the evening and early morning. On average each cow yields four gallons daily.

In the stable are two ponies and a foal, along with six Hereford and Aberdeen Angus calves. The latter are being bred for meat. There are also eleven Friesian calves and twelve heifers, or cows that have not yet calved. The feeding stocks are kept in the large new barns. These are scenes of activity when being stocked with hay and grain after the hay-making and harvest in July and August. Another group of buildings houses the tractor, reapers,

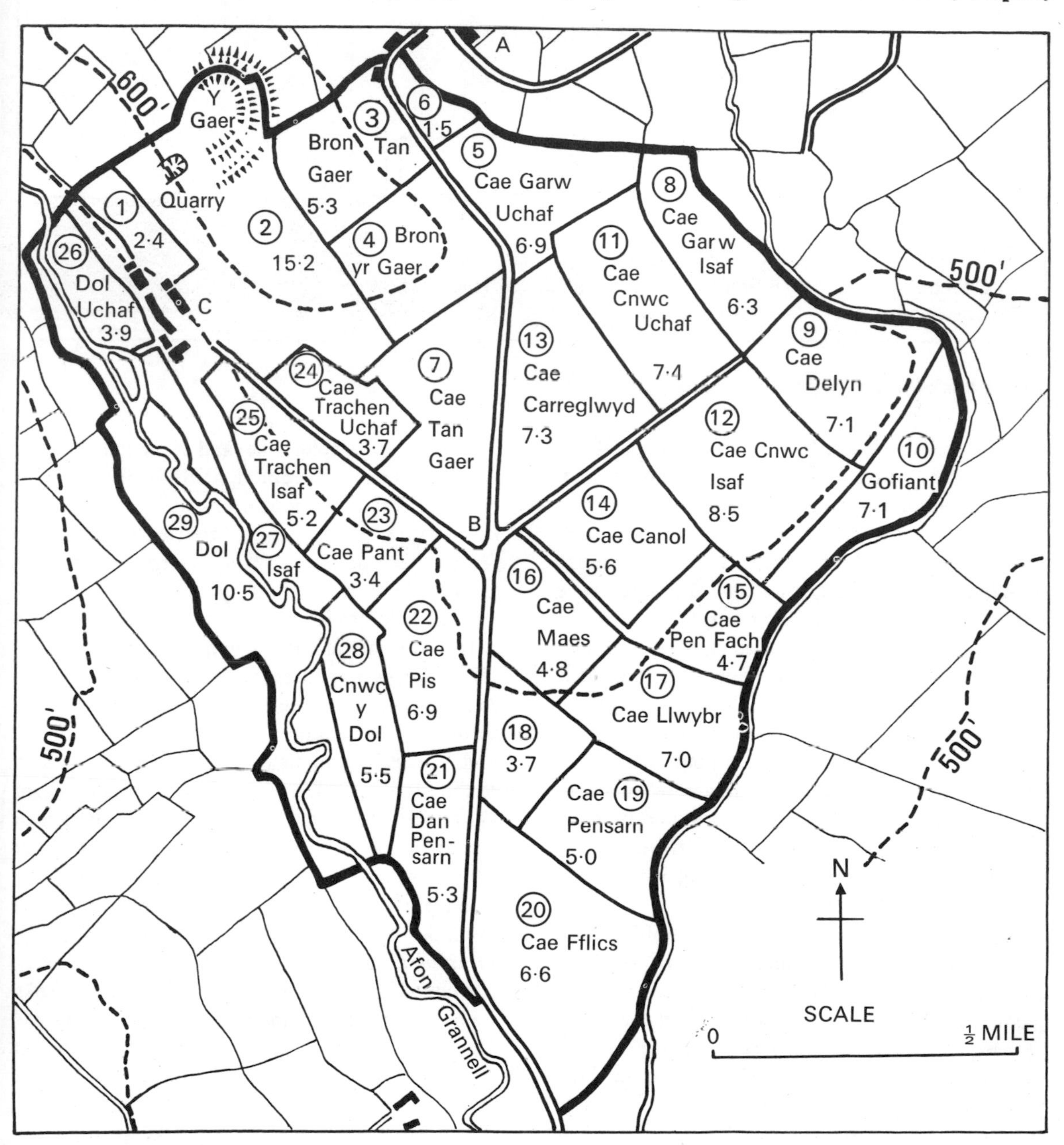

12. Map of Maes Mynach Farm

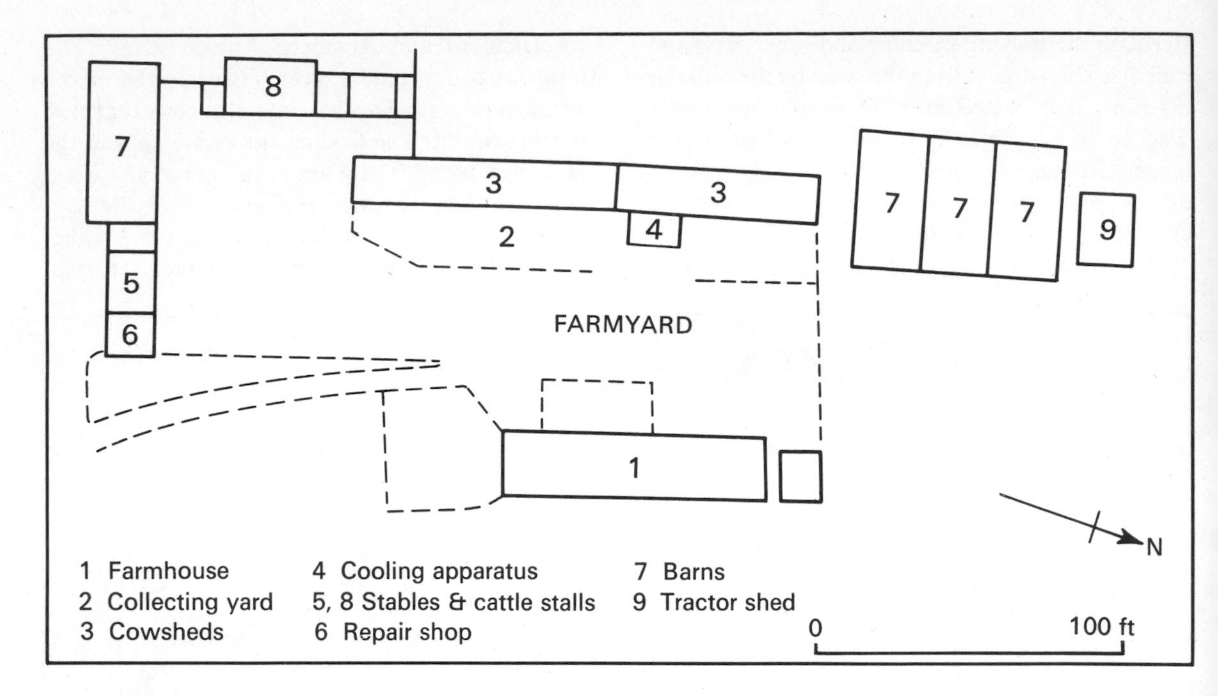

13. Plan of Maes Mynach Farm

binders and the Land Rover. Mr Davies and his sons make running repairs to the machinery in a small repair shop.

A Land Rover took us up the hill which shelters the farm from north winds (Fig. 12). Near the top, the car crossed a ditch and a steep bank which were part of the ancient walls of a Gaer, the centre of a Welsh community which lived 2,000 years ago. From this spot an extensive view of the farm can be seen. On the sides of the hill, rock is near the surface and 200 sheep graze on rough grass.

The main part of the farm is bounded by two streams. Afon Grannell, the larger of the two, winds along a strip of damp meadow and pasture. Both streams are very useful because the cattle can get water on either side of the road which runs across the middle of the farmland.

The land with deep, well-drained soil lies on a low platform between the valleys of the two streams. Grass, the chief crop, provides hay and pasture for the stock. It grows well in the damp climate and is regularly re-seeded. All the fields in this part of the farm are ploughed for two or three years and sown with either barley or mixed corn (i.e., a mixture of barley and oats); they are then reseeded with grass which is allowed to remain for five to ten years before the land is ploughed again. A part of one of the arable fields is planted with potatoes and other root crops and, occasionally, kale—a variety of cabbage used as cattle fodder. The farmer grows on his own land as large a proportion as possible of the food for his cattle.

Each field has its own name, and some of the names are part of the history of the farm. "Pen-sarn", for example, means the end of an ancient road, in this case the straight road which goes through the middle of the farm land. "Cae" and "Maes" mean field. "Maes Mynach" refers to the Monks' field. The monks occupied the farm about 600 years ago, when the farm belonged to Strata Florida Abbey. "Dol" is a meadow, "Cae Delyn" may be so called because it is shaped like a harp (telyn), and "Garw" refers to rough ground. "Gofiant" was originally "Gofaint", meaning a smithy.

Many families in West Wales can trace links with their farms for many generations back. Mr Davies' ancestors farmed Maes Mynach for at least seven generations, and Mrs Davies has an equally long connection with a neighbouring farm. There are also records of the sons who left the farm to become doctors, preachers and teachers.

Pride in past history and present progress go together on this farm.

Most of the land in the western part of Cardiganshire is less then 600 feet high and farming is not as difficult as on the uplands of Central Wales. The farms are scattered in the depth of the countryside and although there are some small hamlets, there are few villages. The occupants of neighbouring farms are often related to one another. Such farms are found in most western parts of Wales from North Pembrokeshire to the Llŷn Peninsula and Anglesey. Within this area a high proportion of people speak Welsh.

Exercises 1. How many gallons of milk are drunk daily in your school? Where does the milk come from? How large a herd would be necessary to produce this quantity?

2. The following are records of the use of land on Maes Mynach farm for the years 1962–64. Draw maps to show the use of the land in any one year, colouring grasses in shades of green and crops other than grasses in shades of brown. The field numbers are given in Fig. 12.

Field Number	*Crop in 1962*	*1963*	*1964*
1, 2, 3 and 4	GL	GL	GL
5	G	GH	G
6, 7, 8, 9, 10 and 11	G	G	G
12	G	B	G
13	G	GH	GH
14	BR	GH	BR
15	G	BR	B
16	G	B	BR
17	G	B	GH
18	G	G	G
19	B	G	G
20	B	GH	G
21, 22 and 23	GH	GH	B
24	G	G	G
25	GH	GH	GH
26, 27, 28 and 29	PG	PG	PG

GL—Grass ley reseeded about every ten years. G—Grass for pasture, in rotation or occasionally ploughed. GH—Grass for hay, in rotation. B—Barley, or mixed corn. R—Root crops. PG—Permanent grass.

3. Draw a simple sketch map of Maes Mynach farm showing the two streams and the farmhouse. Divide the farm into three separate areas, in each of which the land is used differently. Explain why the land is used in this way.
4. Draw a column to represent in length the total acreage of the farm. Subdivide the column to show the averages for 1962, 1963 and 1964 of (a) permanent grass, (b) grass for pasture, (c) grass for hay, (d) grain and other crops. What does the diagram tell you about the kind of farming carried on?
5. Describe the work of three people connected with the production and transport of milk.
6. Write an account of a farm in an area you know well. Draw a sketch map of the farm or, using a six-inch Ordnance Survey map, draw a land use map of the farm showing the crops grown. Find out about the crop rotations used on the farm.
7. (For readers with a knowledge of Welsh.) What do the field names tell you about the farm?
8. Draw a diagram of the farm buildings and write short notes on it explaining how the buildings are used.

5 | NEUADD—A BRECONSHIRE SHEEP FARM

On the great escarpment of the Black Mountains in Breconshire, the farmer and shepherd of Neuadd farm with their four dogs round up sheep for the annual shearing in early June. The mountain covers over twenty square miles and, as this is common grazing land shared by many farms on its margin, there are over 50,000 sheep on it (Fig. 14). From these, the farmer collects the 2,000 which belong to his own flock, identifying them by means of coloured markings on their backs. The job requires skill and patience on the part of men and dogs; any unnecessary movements on their part scatter the sheep. As they move down the scarp, the sheep are concentrated on the lower slope leading into the farm. The arrow leading from X on Fig. 14 shows the route they follow.

When ready for shearing, some of the sheep are driven into the farmyard and then through a door (marked A on Fig. 15) into the shearing shed. This door is in two halves so that the top half can be opened without letting the sheep escape. There are four shearers, the farmer, his two sons and another shearer, each operating a clipper on the end of a cable driven by a belt from a diesel motor. They remove the fleece in one piece, roll it up and tie it with a strand of wool. The shearers work fast and the sons, who practise a New Zealand method of shearing, are able to go faster than their father, who is himself a former champion shearer. (Some young Welshmen go to New Zealand each year to take part in the shearing there.) At the end of the shearing the fleeces are taken away by truck to be sold. The sheep come out at B (Fig. 15) and are returned to the mountain.

In August the sheep are rounded up again for washing, the purpose of which is to prevent scab caused by mites in the fleece. Washing used to be compulsory by law and was usually attended by a local policeman, but in many counties, including Breconshire, this is no longer so because the disease has almost died out. Mr Stephens, the owner of the farm, continues to dip his sheep as a precaution; he also considers that it improves the quality of the fleece.

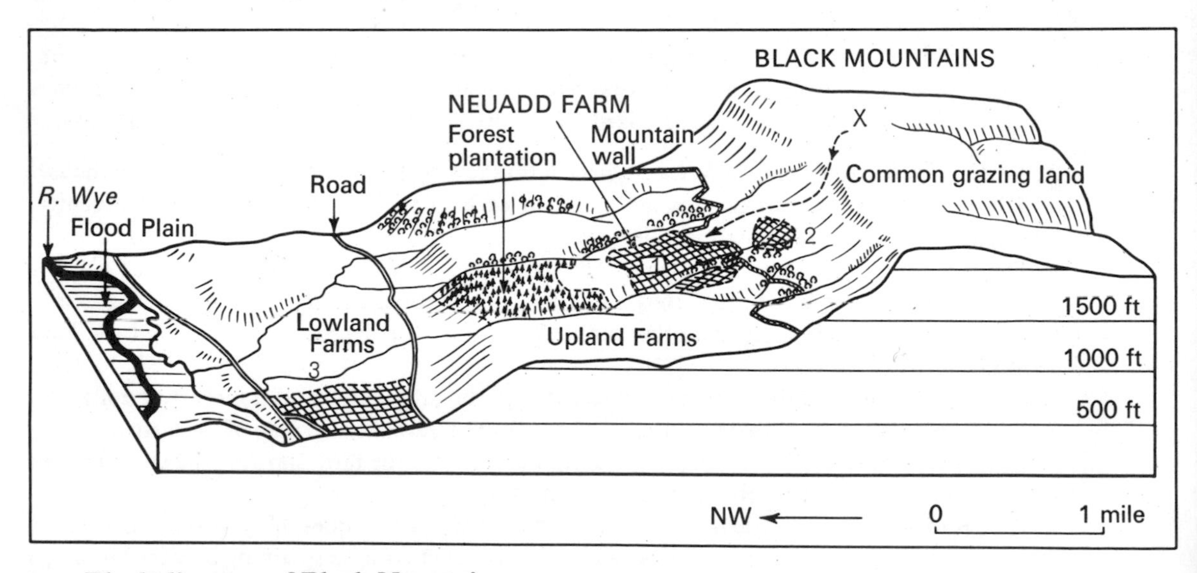

14. Block diagram of Black Mountains

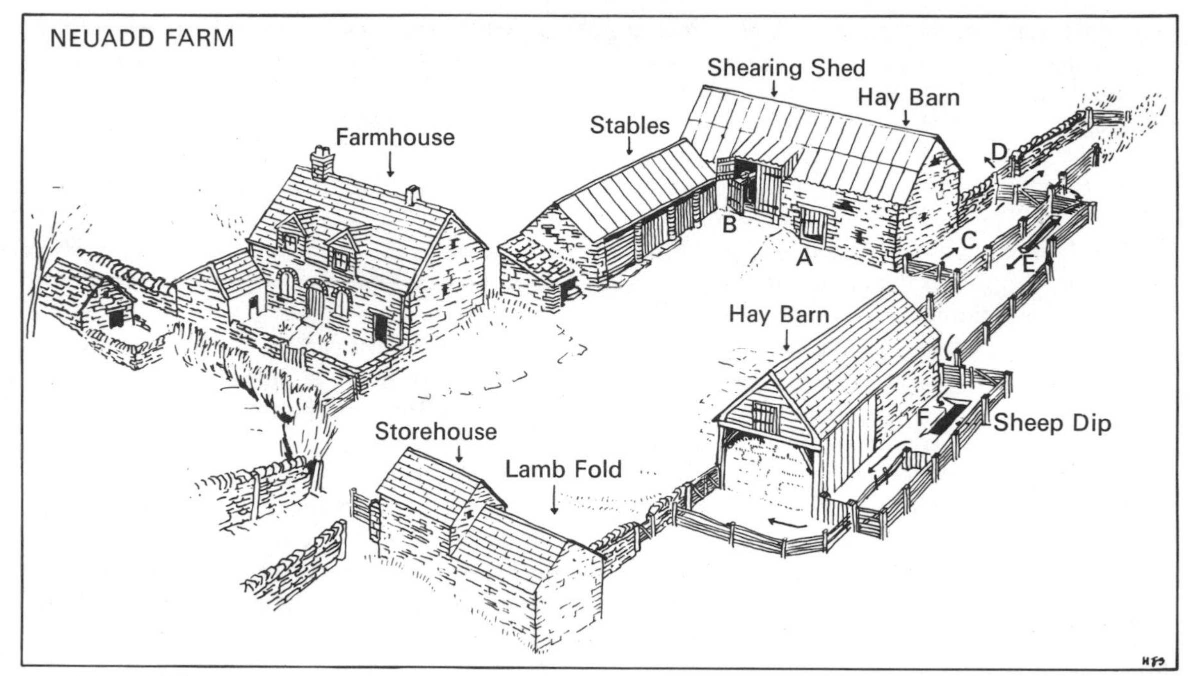

15. Neuadd Farm

At dipping time the sheep pass from the farmyard through the fenced enclosures (shown at C, Fig. 15). The few strays belonging to other flocks are separated at D. At E the sheep walk through a shallow tank containing a solution of copper sulphate to prevent footrot. To lure them into the main dip, five other sheep are kept as decoys alongside (at F).

In September the lambs are weaned by separating them from their mothers for a few weeks, during which time they graze in fields on the detached part of the farm (numbered 2 on Fig. 14), before being sent to the mountain. In September, too, the annual sheep sales take place in the nearby towns of Hay and Talgarth. Mr Stephens is not much concerned with the sales because, in conjunction with his brother, he runs a lowland farm (marked 3 on Fig. 14). There he fattens his own sheep and sells them direct to the meat trade. But most of the local hill farmers sell store lambs at the sales, to be fattened in other areas such as the Midlands of England. About 15,000 sheep are sold every year in Hay and about 10,000 in Talgarth.

In October about 200 of the lambs are collected to be "tacked" or sent by lorry to another farm in Monmouthshire where they spend the winter. These return to the mountain in the following April. The western coastal districts of Wales are also a favourite wintering area, but since they now concentrate on dairy farming, they no longer take as many sheep as formerly. About 500 lambs are also wintered on the lowland farm. Most of the ewes belonging to the Welsh Mountain breed remain on the mountain during the winter.

At Neuadd, about ten acres of swedes, mangolds and rape are grown to help feed the stock in winter, as well as hay which is cut in July and stored in the barns. The root crops are fed to the animals in the field.

During severe weather on the mountains in winter the sheep must be fed and, if necessary, dug out of the snow. By the beginning of April lambing has started. On average 22 lambs are born to every 20 ewes. Some of the newcomers replace the older ewes and themselves bear lambs in due course while the male lambs are sold later in the year.

So far, only the main events of the year have been mentioned, but the shepherd has work to do every day, examining the condition of the sheep and supervising their grazing so that they do not congregate too much in one place or too near the farm. He keeps a particularly careful watch during the lambing season when foxes are a danger.

Neuadd and the lowland farm are worked in combination. The shepherd now lives in what used to be the farmer's house at Neuadd (see Fig. 15) and the farmers themselves live on the lower farm near Talgarth. Both farmhouses are modernized and have electricity. Neuadd has 134 acres of enclosed land. The lowland farm of 400 acres has 150 cattle and also fattens the Radnor and Suffolk breeds of sheep brought down from Neuadd. So there is much movement of sheep from the uplands to the lowlands. Grain and root crops provide feeding stuffs.

Most successful sheep farms in this part of Wales have the following advantages:

1. An extensive mountain or moorland area where the sheep can be grazed most of the year.

2. An enclosed area near the mountain where sections of the flock can be kept at certain times. These fields (numbered 2 on Fig. 14) are known in many parts of Wales as *ffriddoedd.*

3. A main farm near the edge of the moorland providing winter grazing for part of the flock.

4. Farm buildings arranged so that sheep can be easily gathered for shearing and dipping.

5. Road access so that regular movements of sheep to and from the lowlands can be carried out.

As in many other parts of Wales, the differences in height correspond to differences in the quality of the land. Here, on the escarpment of the Black Mountains, there is high moorland over 1,800 feet, a shelf of upland between 1,000 and 1,500, and the valley floor below 500 feet. The soils are poorer and the climate wetter and colder at the higher than at the lower levels. This is the reason for the movements of sheep between uplands and lowlands, since the mountain land is too poor to fatten lambs and the upland farms cannot grow enough winter food for all the stock.

Exercises

1. Draw a section from the Black Mountains to the river Wye, using as a guide the front edge of Fig. 14. Show, as nearly as possible at their right levels, the mountain wall, Neuadd farm, the forest and a lowland farm. Add arrows showing movements of sheep at different times of the year.
2. Make a calendar showing the main events and activities on a sheep farm in one year.
3. Draw a simple plan of Neuadd farm based on Fig. 15. The farm is approximately 80 yards long and 40 yards wide. Add brief notes on the purpose of each building and the materials used.
4. Fig. 14 shows distinct areas of land at three different levels: (a) what approximately are the heights of the levels? (b) describe briefly the physical character and agricultural use of the land at each level.
5. Summarize the main differences between the dairy farm (Maes Mynach) and the sheep farm (Neuadd) under the headings: (a) size of farm and type of land on which it is situated; (b) crops, stock and use of farm buildings; (c) daily and annual routine on the farm.
6. Describe what happens on a sheep farm when (a) shearing and (b) dipping are carried out.
7. Write an account of the work of a shepherd.
8. Make a plan of a farm which you know and show what the buildings are used for.
9. Why is it important in a farming area: (a) to prevent dogs from chasing sheep; (b) not to break hedges or fences; (c) to shut gates?

6 | ABERGAVENNY—A MARKET TOWN

The sketch map (Fig. 16) shows that Abergavenny stands where the river Usk emerges from a deep valley between the Black Mountains on one side and the steep edge of the South Wales coalfield on the other. It is a meeting-place of roads on the border between highland and lowland.

Tuesday is market day in Abergavenny. From early in the morning Land Rovers and trucks loaded with animals come into the town from farms and villages within a distance of about 20 miles. The main street is narrow and congested in places in spite of one-way traffic. In the market, sheep are unloaded into pens and handed over to the care of an auctioneer who sells them on behalf of the farmers. There is a toll charge for each animal, payable at the market superintendent's office. As the pens fill up, farmers and dealers walk round and examine the sheep (see Fig. 17). Later, the auctioneer, standing on planks over the pens, will begin to sell batches of sheep, taking bids so quickly that it is difficult for the ordinary spectator to keep count.

In a covered sale ring, cattle are auctioned, mostly Hereford breeds; store calves from farms in the hill districts are sold to farmers in the lowlands for fattening. Fatstock for the slaughterhouse and butchers' shops is also sold. In other parts of the market are covered enclosures for pigs and poultry, and displays of tractors and agricultural machinery, seeds and fertilizers. Wool buyers and barn erectors maintain offices in the market. One part is reserved for the sale of ponies, now much in demand because riding schools and pony trekking have become popular.

A little way from the stock market and next to the Town Hall is the Butter Market, a large covered hall with about fifty trestle tables laid out. Although butter used to be one of the main products sold here, the tables are now covered with fruit and vegetables, clothes and toys.

The farmers also have business outside the market and the town caters for this. They may have

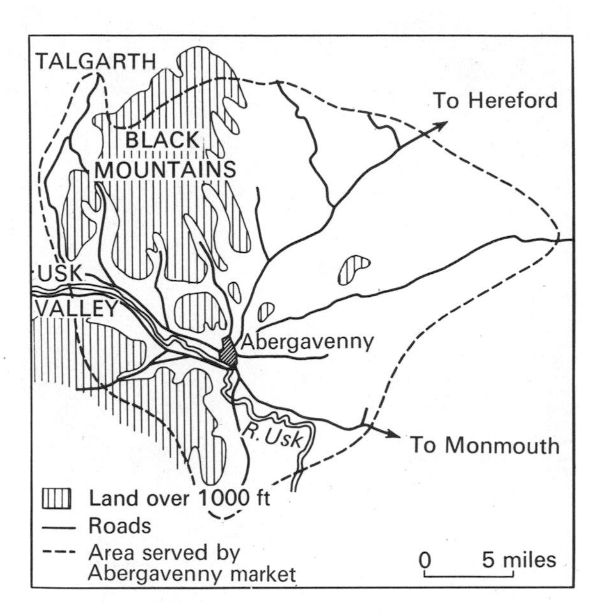

16. Situation of Abergavenny

dealings with estate agents and auctioneers whose offices are usually near the market. They need to visit banks. Dealers buying stock may remain overnight in Abergavenny and a number of hotels, public houses and cafés supply their needs. As most people come in from the surrounding countryside by car, ample parking space and garages are necessary. Another essential person in every market town is a veterinary surgeon.

Fig. 18 shows the use which is made of the land and buildings in the centre of the town. The drawing is a simplified version of the Ordnance Survey map on a scale of twenty-five inches to a mile on which all buildings, walls, fences and streets are shown. Each type of property has been given a different letter according to the following system:

R —residences, or houses in which people live.

S —shops and other buildings in which things are sold.

17. Photograph of Abergavenny Market

W—warehouses or yards in which goods are stored, including builders' yards.

T—all buildings and land concerned with transport such as bus stations, railways, car parks, garages, etc.

O—offices and banks.

M—factories.

P—public buildings such as schools, churches and libraries.

F—farm land, fields.

H—hotels, cafés and public houses.

One can walk through the centre of an old town such as Abergavenny and learn much of its history and geography. Starting at 1 (Fig. 18) on the banks of the river Usk, it can be seen that the castle and much of the town were built on high ground well above the level of the fields near the river, which is liable to flood.

The route now crosses the river Gavenny, the confluence of which with the Usk gives the town its name. Not far up the Gavenny from its confluence is a small waterfall and near it an old mill, now a welding and repair shop. At 2 is a residential area of rather large Edwardian houses, i.e., built in the early years of this century, near the railway station. Mill Street (3), in contrast, is a narrow street of small houses where the old town began. At 4 there was once a woollen mill driven by water power from the mill stream. On a map made about 1800, the space now occupied by the bus station was called Tenter field, and the cloth from the mill was laid out there to stretch and bleach. The mill is now a woolstapler's warehouse used for storing wool. A neighbouring house was formerly the tannery which made leather from skins of animals sold in the town.

At the beginning of Cross Street (5), are several hotels and cafés near the bus station and car park.

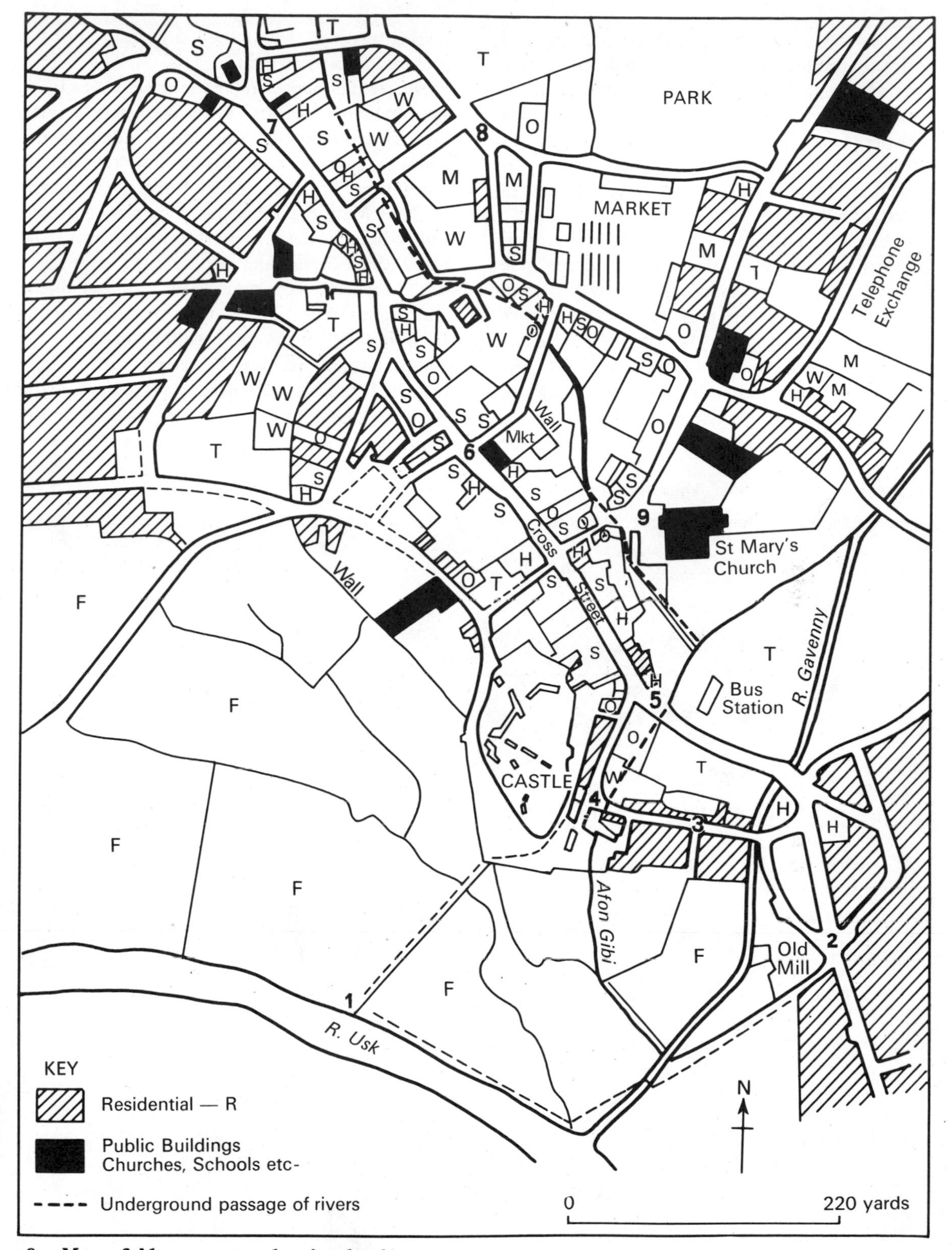

18. Map of Abergavenny showing land use

Find Cross Street on the photograph (Fig. 19), and notice the buildings in it. Many of them are over 200 years old. Two of the hotels which may be recognized from the map (Fig. 18), catered for travellers in coaching days when Abergavenny was a stage on the main route through Wales. At the rear of the hotels were stables where the horses rested overnight.

Near the Town Hall (6) the main street narrows, but this is, nevertheless, the busiest shopping area. Along the side streets, shops soon give way to small offices or houses. A large part of the old town, once a maze of small streets, has been demolished and is used for parking cars until it can be rebuilt. The street names are a clue to the former activities of the town, e.g., Flannel Street. Further along the main street is Frogmore Street (7), where a cattle market was held during the last century until the present market was built.

Nearer the market, there are fewer shops to be seen and more industry (8)—a toy factory, a small foundry and an egg-packing station. Adjoining the market are several estate agents and other offices. The journey finishes in Monk Street at the church of Saint Mary (9). This was once part of an old priory built in the thirteenth century, hence the name Monk Street, and next to it is a large stone tithe barn where the grain, paid as a levy to the church, was stored. Tithes are no longer paid in this form, but the barn is still used for storage.

Abergavenny is clearly a town with a long history. It began as a Roman fort placed around the present Cross Street. The castle was built by the Normans some thirty years after their conquest of

19. Air photograph of Abergavenny

England. The town which grew up around the castle was later walled to protect it against Welsh raids from the surrounding hills. The line of the wall, little of which can be seen today, is shown on Fig. 18. In subsequent centuries Welsh people settled in the town and the market grew in importance. Industries like woollen manufacturing, and especially flannel-making, developed at the watermills. Control of the route into Wales gave Abergavenny a coaching trade, but it never became a main railway centre because no railway was built far up the Usk valley. Today, the Heads of the Valleys Road begins just south of Abergavenny and carries a heavy through traffic into the South Wales coalfield.

Many of the older towns of Wales have had a similar history. Market towns and route centres like Brecon, Builth, Conway and Caernarvon grew up in the shadow of castles and a few, like Cardiff and Neath, have expanded very considerably. The growth of towns in South Wales and the eastern border followed the Norman conquest. In the more resistant areas of North Wales the castle towns were not built until the conquest by Edward I in the thirteenth century. A few towns, like Tregaron, were of purely Welsh origin.

The market towns scattered throughout Wales have played a vital part in the sale of agricultural produce. The products of dairy farming are now processed through creameries, and the market towns are mainly concerned with the sale of stock. Although some stock farmers have begun to supply stock directly to dealers, market towns like Abergavenny will continue to have a useful purpose in serving the country districts of which they are the centre.

Exercises

1. Draw a sketch map of the central part of Abergavenny. Name or initial the castle, church, market and town hall. Colour distinctively the residential areas, shops, warehouses, transport premises, offices and banks, factories and hotels. Colour fields and parks green. What does the map tell you about the town? Why are there no buildings on the land near the river Usk?
2. Make a simple plan of the main streets of Abergavenny. Draw circles or ovals around those parts of the town which are mainly concerned with (a) manufacture and storage of goods, (b) transport, (c) hotels and shops, (d) offices. Explain, as far as possible, the positions and shapes of these areas.
3. Draw a simple sketch map of a market town you know well or have visited, and write a brief account of its market.
4. Using the scheme of mapping suggested on page 23, make a map of the town or part of the town where you live, showing the use of buildings and land. (This can be done as a class activity.) What does the map tell you about your town? What is its most important activity? Are shops, offices, factories, etc. located in particular parts of the town. Are some parts much older than others?
5. With the aid of Fig. 16, describe the position of Abergavenny and the area served by its market.
6. Make a list of as many market towns in Wales as you can find. Find their positions from your atlas. How many of them have castles?
7. What do buildings and street names tell you about: (a) the market, (b) the history of the town? Answer this question with reference to either Abergavenny or another market town you know well.
8. Compare Fig. 18 with Fig. 19. Make a simple sketch based on Fig. 19 and add notes to your diagram explaining features of interest.

7 | COAL MINING IN NEW TREDEGAR AND ABERNANT

The photograph (Fig. 20) shows a typical mining town, New Tredegar, in the Rhymney valley. The long rows of terraced houses follow the direction of the valley and the town itself is long and narrow (see also Map C). The houses are made of Pennant Sandstone, a locally quarried rock which weathers dark brown with age and which, because of its toughness, forms the high ground between the valleys. The town lies in one of the deep valleys which dissect the plateau; near the bottom of the valley is the Elliot coal mine.

A closer look at the colliery in the picture shows the electrically driven winding wheels which lift the cages carrying men and coal from below. The engine-house is alongside it and to the left are the screens where the coal is divided into fine coal for power stations and larger pieces for coking and house coal. In the background of the photograph is the waste tip to which shale and other waste rock from the workings are carried by overhead ropeway. The tip represents the waste removed from the mine since it started in 1890. In 1967 Elliot colliery was closed, like a number of other mines of about the same age.

The block diagram (Fig. 21) shows what lies underneath the valley. The shafts of the colliery descend to the base of the Lower Coal Measures, nearly 1,000 feet below sea level. The Middle and

20. Photograph of New Tredegar

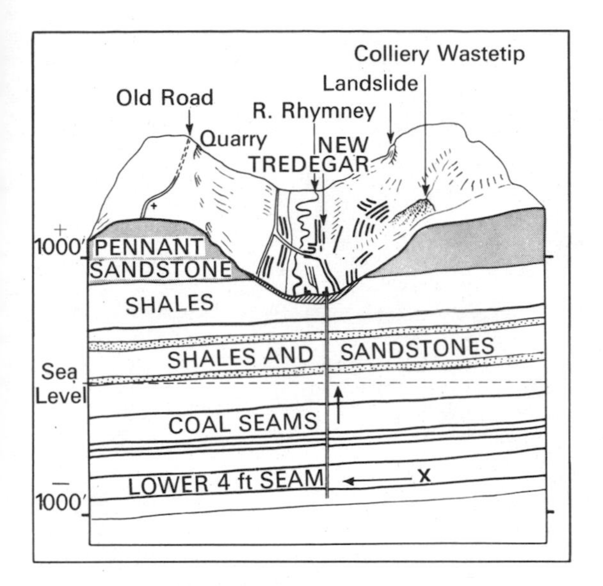

21. Block diagram of Rhymney valley

Lower Coal Measures contain more productive coal seams than the Upper Coal Measures which include the Pennant Sandstones. Workable coal amounts to less than one twentieth of the thickness of rock; the rest is clay, shale and sandstone. The rock layers dip or slope gently from east to west. Most of the workable seams were between two and six feet thick. At Elliot colliery five main seams were worked, the last before closure being the deepest, namely the Lower Four Foot seam.

The seams were formerly mined by colliers undercutting the coal with picks and then firing explosive shots near the top of the seam to bring the coal down, so enabling it to be worked by hand shovel. Today, about 60 % of the coal is cut and loaded by machinery. Fig. 22 shows one type of machine, a trepanner, extracting the coal and loading it on to a conveyor belt. Above the coal seam the rock strata "give" slightly after the coal has been cut and are then held, either by pit props, the old method, or, as in Fig. 22, by hydraulic jacks. After one shift of miners has removed the coal along the length of the coalface for several feet, the roof support and conveyor belts are moved forward before the next shift starts cutting. From the conveyor belt the coal is transported to the bottom of the shaft in small trucks.

The mining industry has changed considerably in the lifetime of older miners working today. No longer do colliers return from the mines covered with coal dust. They go to work by car or bus and there are pithead baths at every mine. The outward appearance of the towns has not changed much except that the newer housing estates spread on to higher land above the town. No picture of a mining town would be complete without mention-

22. Trepanner at Work

23. Abernant Colliery

ing the many stone chapels, the social clubs and the rugby ground, the latter usually on the limited area of flat ground near the river.

Before the mining towns grew up, the wooded valleys of South Wales were used for farming. Near New Tredegar, the best farms and older roads were on the uplands where they still exist. Those farms which were in the valley have been covered by the growth of the town. Capel y Brithdir, the oldest centre of the community, is not a nonconformist chapel like those in the town but an ancient church near which stood a Celtic stone monument, now in the National Museum of Wales.

Abernant—an anthracite mine

In contrast with Elliot colliery is the new anthracite mine at Abernant (Fig. 23) in the Cwmgors valley, about ten miles north-north-east of Swansea. The white towers shown in the photograph contain the electrically driven winding gear and alongside the mine is a coal preparation plant. No town has grown up alongside the mine as in the case of New Tredegar; Cwmgors, the nearest village, is one and a half miles away. Workers travel daily from the older mining districts in the western parts of the coalfield where anthracite mines have closed down.

If you live in the smokeless zone of a large city, the coal you burn may be anthracite, which is hard and shiny. On burning, it emits no smoke and leaves very little dust. Practically all of the anthracite mined in Britain comes from the north-western part of the South Wales coalfield.

The old anthracite mines near the north-west edge of the coalfield are levels or drift mines. The coal seams dip very steeply and are soon deep below the surface. Most of these mines have already obtained nearly all the more accessible coal and the miners have to travel a long way from the outlet to get to the coal face. In the centre of the anthracite field the coal is so deep that hardly any has been worked until recently. To work these reserves, the National Coal Board has sunk new mines, one of which is Abernant and another Cynheidre in the Gwendraeth valley.

Map A: scale 1: 25,000

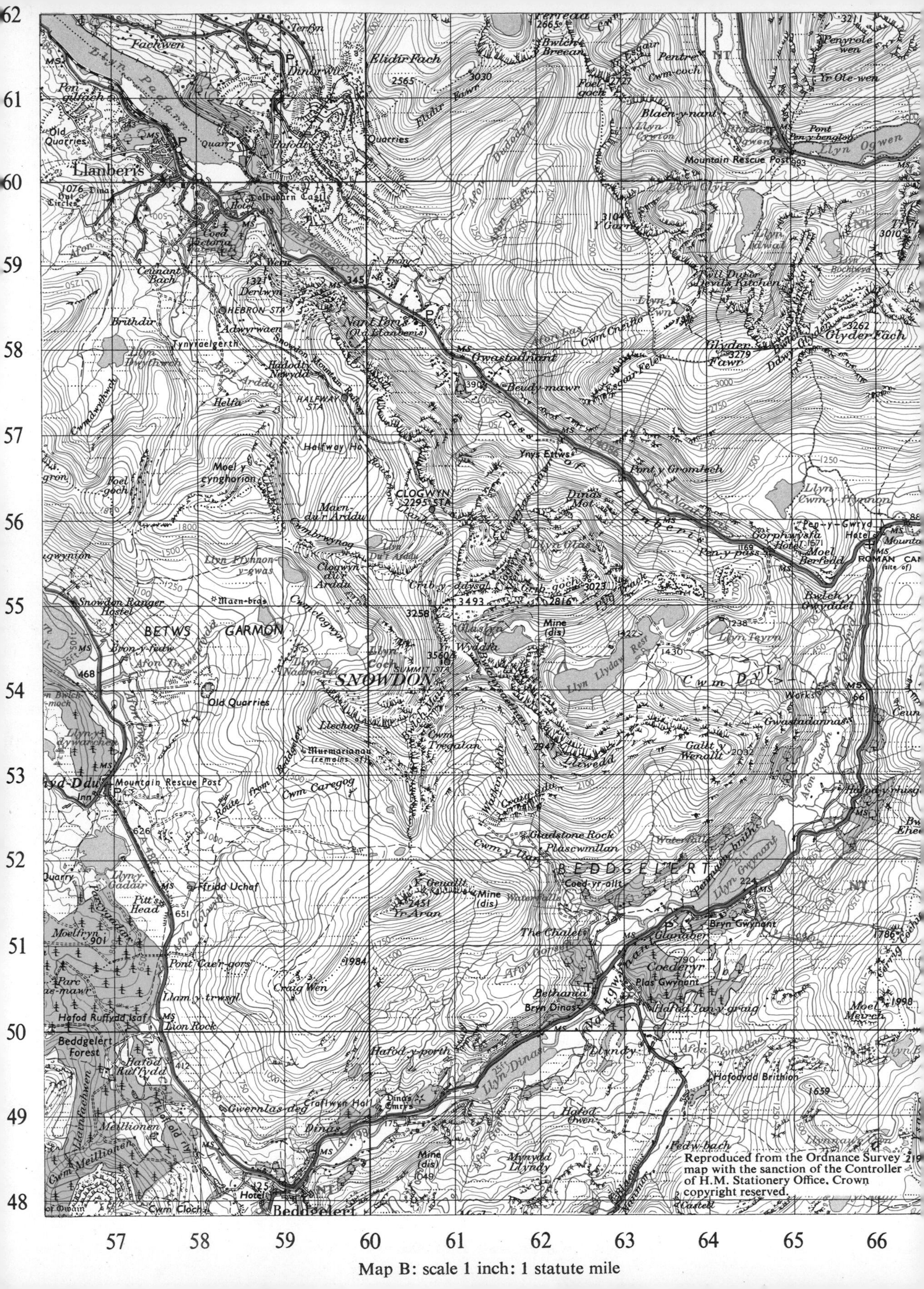

Reproduced from the Ordnance Survey map with the sanction of the Controller of H.M. Stationery Office. Crown copyright reserved.

Map B: scale 1 inch: 1 statute mile

One of the best seams of anthracite is the Peacock seam which comes to the surface near Brynamman. Less than four miles to the south of its surface outcrop it is 2,400 feet deep. The new mine at Abernant was sunk to work the seam at that depth. When, after a few years' work, the shafts were finished and the mining engineers started to drive the headings, that is, to excavate the tunnels to the workings, they found the pressure of the rocks to be so great that it forced in the sides of the tunnels. For the time being, until the problem can be solved, the Red Vein only is being worked, at a depth of 1,200 feet.

The limits of the coal to be worked at Abernant are defined by two great faults or breaks in the continuity of the rock strata. West of the mine, the rocks have subsided 1,100 feet and to the east, 600 feet. To the north, mining is limited by older, shallower mine workings. To the south, the depth of the coal seams descends over 3,600 feet below the surface and it is technically impossible at present to mine at that depth.

The problems of the coal-mining industry are to some extent a result of the nature of the earth's crust. First, although mining engineers take borings and gain a detailed knowledge of the structure of rocks underground, there are always risks that the seams may not behave as anticipated. Secondly, coal must eventually become exhausted. There are still very large reserves of coal in South Wales (2,000 million tons proved to exist and a further 3,000 million estimated), but several of the older mines have become exhausted. The workers must then turn to other mines or other forms of employment. Most of the new developments in the South Wales coalfield have been designed to exploit coal where it has not previously been possible to mine it.

Exercises In answering questions 1 to 6 use Map C.

1. Compare Map C with the photograph (Fig. 20). (a) Where was the photograph taken and in which direction was the camera pointing? (b) Find the Elliot colliery on both map and photograph. How deep is the valley and why is the colliery at the bottom? (c) Find Jubilee Road and Queen's Road. Which Queen and whose Jubilee do they refer to? What does this tell you about the probable age of the town? (d) Find in your atlas New Tredegar and the valley in which it lies.
2. Using Map C, draw a simple sketch map of the mining town showing in distinct shading or colours: (a) land over 1000 feet high; (b) the river Rhymney; (c) the position of the coal mine; (d) the shape of the town.
3. (a) Draw a section across the valley from Tyr-Capel to Tŵyn y Gwynt. The horizontal scale should be the same as that of the map, and the vertical scale one inch to 500 feet; (b) show on the section the coal mine and upper part of the shaft, the river Rhymney, the town and the woods.
4. Compare Map C with the photograph (Fig. 20). (a) Find on the map the railway, the colliery, the waste tip and the woodland; (b) on the photograph find Phillip's Town and Elliot's Town; (c) why was Phillip's Town built where it is?
5. (a) What kind of rock is found in the quarries and what has it been used for? (b) What was the purpose of the old coal levels? (c) What do you think is the purpose of the aerial ropeway? (d) Comment on the routes followed by main roads and railways. (e) What is the significance of the boundary along the river Rhymney?
6. Where is the farm land? Find some examples of farm houses and state where they are situated.
7. Compare mining at Elliot colliery with that at Abernant. Use the following headings for your answer: the type of coal; the age of the mine; the depth and dip of the coal seams; the methods and problems of mining.
8. What is anthracite? In which part of the South Wales coalfield is anthracite mined? Why is it difficult to mine? What are its uses?
9. Describe the main stages in the mining of coal.

8 THE TAFF VALLEY

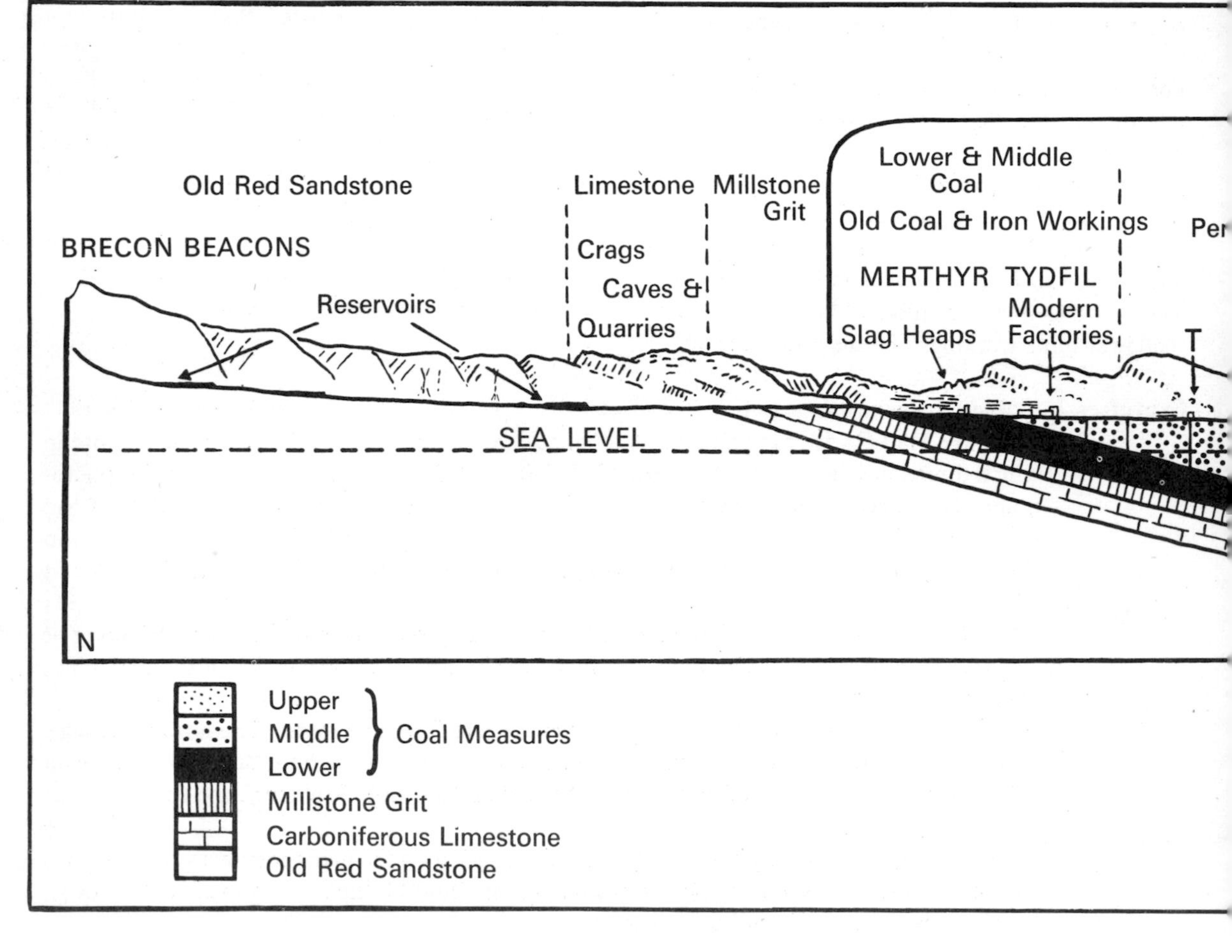

Eight main rivers cross the South Wales coalfield in deeply cut valleys. One of the best known is the Taff, the largest tributary of which is the Rhondda. The source of the Taff is among the high, desolate mountains of the Brecon Beacons where, in its upper reaches, the river flows across Old Red Sandstone rock (Fig. 24). The river then crosses a belt of limestone in a narrow valley with steep, grey crags and quarries. At Merthyr Tydfil it reaches the northern edge of the coalfield.

Here the Lower Coal Measure rocks outcrop or come to the surface and there is hardly an acre of land that has not been disturbed. Embedded in shales are small round balls or nodules of iron ore from an inch to four inches in diameter. Even before the eighteenth and nineteenth centuries iron miners dug out the iron ore from shallow pits or small levels driven into the hillside. Then they washed the ore in streams, or in water released from small ponds, to clean away the clay and sand from the heavy iron. In the iron works the ore was smelted by using charcoal obtained from the woods which covered the hillsides. The bars of iron were carried over the hills to the coast on the backs of

24. **Sectional diagram of Taff valley**

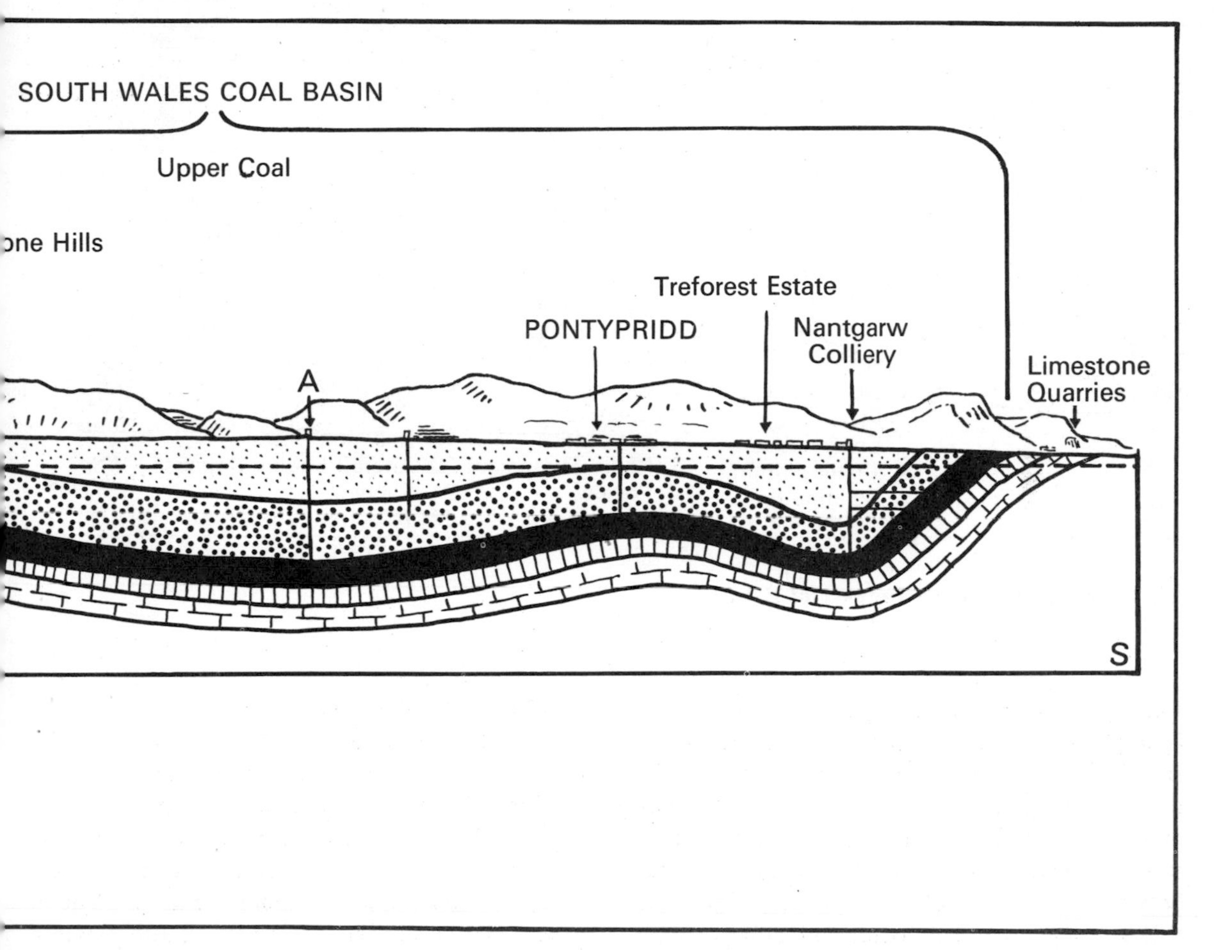

mules or horses. When the canal was built from Cardiff to Merthyr in 1794, transport in bulk became easier.

By the early nineteenth century the Cyfarthfa and Dowlais works had become two of the greatest iron works in the world and in 1830 Merthyr was the largest town in Wales, rivalled only by Swansea. The Bessemer process for making cheap steel, started in 1856, enabled Merthyr to produce rails for the American transcontinental railways, but foreign ores now had to be imported and brought up the Taff valley by rail. This raised the cost of the ore to inland centres. Swansea and the coastal areas therefore gained an advantage over Merthyr and the other iron and steel manufacturing towns in the northern part of the coalfield. Today, the remains of the great iron works can be seen only in the ruined furnaces and tall slag heaps around Merthyr. The present industries of Merthyr have little connection with coal or iron. Modern factories like Hoover's and Kayser Bondor's have been built, the latter on the site of the old Dowlais iron works.

The rock strata in the South Wales coalfield have been folded into the form of a basin (see

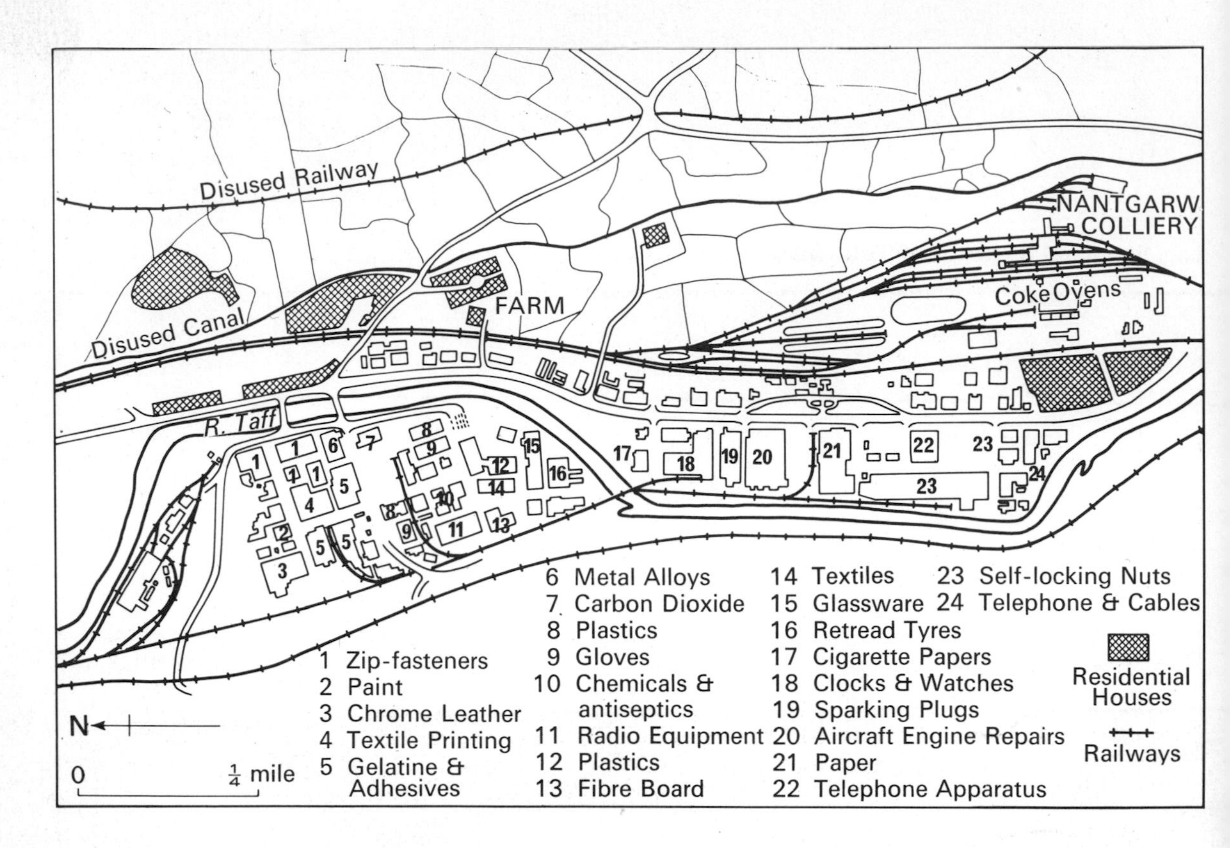

25. Plan of Treforest Industrial Estate

Fig. 24). At Merthyr, they dip gently into the basin from the north side, but rise sharply on the south side. There are also some minor upfolds within the basin. The earliest mining took place on the margins of the coalfield because the coal seams there were at or near the surface. Between 1850 and 1914, however, deep mining for coal developed towards the centre of the coal basin, and railways were built along the valleys. Long ribbon-like towns filled the Rhondda and middle Taff valleys, which had previously been rural. Fig. 24 shows that the mines become deeper as one goes south, so that a seam of coal coming to the surface at Merthyr, is 750 feet deep at Troedyrhiw (marked T on Fig. 24), 1,200 feet at Merthyr Vale (MV on Fig. 24) and over 2,000 feet deep near Abercynon (A on Fig. 24). In this middle part of the Taff valley are mining towns similar to New Tredegar, though mining is no longer the only occupation for men or even the main occupation in some towns in the Taff and Rhondda valleys.

The Rhondda river joins the Taff at Pontypridd; down the valley from this point is the Treforest Industrial Estate (Figs. 25 and 26), built in 1936 to provide work for large numbers of unemployed miners. Today the Treforest Industrial Estate employs over 7,000 men, more workers than in all the coal mines in the Taff valley. It also provides jobs for over 3,000 women. The busy main road to Cardiff passes through the estate and the railway is alongside it.

The factories at Treforest are rented; Fig. 25 shows the layout of the factories on the estate. The Industrial Estate provides electricity and gas, and steam generated in its own plant, using water filtered from local streams and the river Taff. The latter can be seen in the foreground of Fig. 26, and the factories, over sixty in number, beyond. A great variety of products is manufactured in these factories, ranging from telephone cables to guitar

26. Air photograph of Treforest Industrial Estate

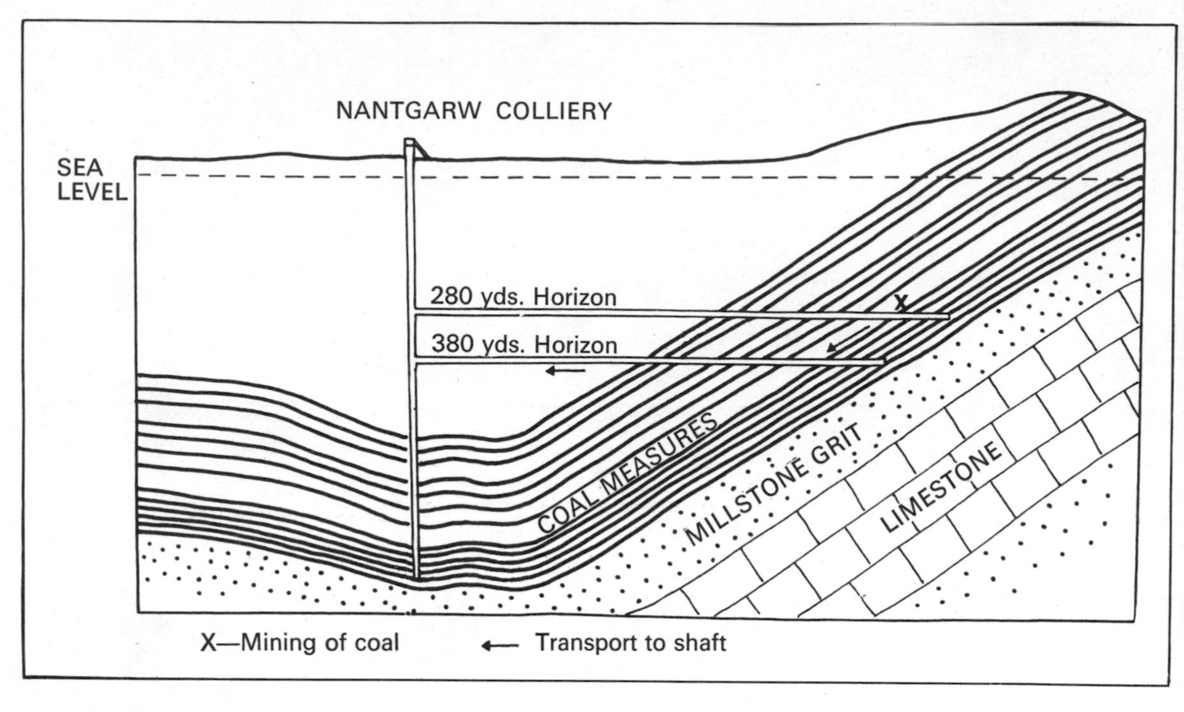

27. Section through workings at Nantgarw Colliery

strings. Usually, there is a shortage of space in the mining valleys, but the Industrial Estate was built in a part of the Taff valley where there was an unusually large unoccupied space in the flood plain, mainly because there were few mines in this part of the coalfield.

Near the southern edge of the coalfield, as shown by Fig. 24, the rock strata, including the coal seams, lie at a steep angle. These seams have always been difficult to mine. In 1946 a new mine was developed at Nantgarw, to work large areas of coal hitherto untouched. A method called horizon mining was used to overcome the difficulty of the steeply dipping coal seams. As Fig. 27 shows, tunnels were driven from the shaft towards the coal at horizons, 280 yards and 380 yards underground. Coal is worked from the upper horizon and carried by conveyor belt down to the lower one. Trains of mine cars drawn by diesel locomotives carry the coal to the bottom of the shaft ready for bringing to the surface. The underground transport system is much better than in the older mines. Next to the colliery at the surface stands a coking and by-products plant. It uses coal from Nantgarw and neighbouring mines and produces 2,500 tons of coke a day for the steel works of South Wales, gas and other products such as ammonia, tar and benzol.

South of Nantgarw the Taff valley narrows at Taff's Well as it passes between the steep escarpments bordering the coalfield. Limestone quarries now succeed coal mines, and from this point to Cardiff and the sea the Taff flows in a shallow valley across the low coastal plain.

The Taff valley is the home of people most of whose ancestors migrated there to become workers in the iron-making and coal-mining industries. Complete dependence on these two heavy industries led to unemployment when they declined. Today the Taff valley still lives partly by coal mining, but it also has a number of light industries scattered throughout the valley as well as concentrated at Treforest. These industries make a wide range of products.

Exercises 1. Using your atlas as well as the information given in this chapter, write an account of a journey up the Taff valley from Cardiff to the Brecon Beacons.

2. Draw a simple section across the South Wales coalfield. Why are the mines deepest towards the centre and south? Why were the older mines mostly on the north side?

3. Contrast Merthyr today with Merthyr in the last century.

4. Write an account of the Treforest Industrial Estate and draw a sketch map to show where it is situated. Why was the estate started?

5. Classify the industries on the Treforest Industrial Estate into about five groups (see Fig. 25).

6. Compare the conditions of mining on the northern margin of the coalfield with those on the southern margin.

7. Illustrate, by reference to the Taff valley, the differences between heavy and light industry.

9 | STEEL

In 1959 a traveller on the express train from Paddington would have seen, after leaving the Severn Tunnel, some miles of straight-edged fields drained by ditches. These were the Gwent Levels, a coastal plain bordering the Bristol Channel and consisting of meadows and pasture land. Today these fields contain the huge Spencer steel works. To walk through the plant from one end to the other, without stopping to look round, takes an hour and a half. Every day thousands of workers arrive there by car and bus from places as far as thirty miles away, many coming from Newport, five miles to the west, and from Cwmbran New Town, a few miles to the north. On this flat land it was possible to lay out the works according to a set plan.

From Fig. 28 and the photograph (Fig. 29), much can be learnt about these works. At the west end, nearer Newport, the raw materials for making steel are stored. The first is iron ore, most of which reaches Newport docks from North Africa, South America and Labrador. Special freight trains, sometimes as many as sixteen a day and each a quarter of a mile long, bring the ore from the docks. About ten thousand tons of ore a day come from Newport docks and a thousand tons a day from mines in Oxfordshire.

Coal is used to make coke for heating the blast furnaces and three thousand tons come daily by rail from the South Wales mining valleys. Four hundred tons of mineral oil a day are also used. Lorries transport daily some two thousand tons of limestone from quarries around the South Wales coalfield.

The material used in greater quantities than any other is water. In its first stage of development the Spencer works was scheduled to consume at least eight and a half million gallons a day, nearly as much as the city of Cardiff uses for drinking and washing. The water is used for raising steam and for cooling purposes; the outer casing and doors of furnaces, for example, are cooled with water. The large cooling towers, which are a prominent feature of the picture (Fig. 29), are necessary to clean and cool the used water, so that it can be re-circulated through the manufacturing processes. Soft water is pumped from the river Usk and hard water from the springs in the Severn Tunnel; both are stored in a reservoir near the works (see Fig. 29).

East of the stockyards are the tall, narrow coke ovens, numbered 2 on the diagram (Fig. 28). When the coal has been heated sufficiently, the operator

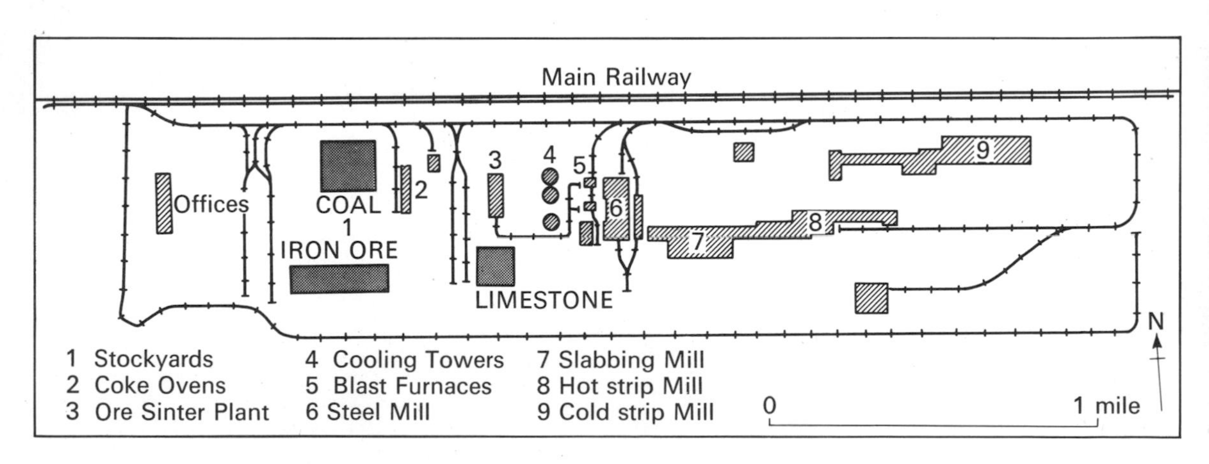

28. Plan of Spencer Steelworks

29. Air photograph of Spencer Steelworks

from his cabin sets in motion an arm which pushes the red-hot coke out of the ovens. It is then cooled by a spray of water. Good coking coal makes a coke strong enough to hold up the heavy weight of ore in a blast furnace. The gas obtained from the heated coal is used as fuel in other parts of the works, and benzol, ammonia and tar are also produced. The iron ore, before it arrives at the blast furnaces, is "sintered" or baked with powdered coke and lime (Fig. 28, 3). This changes it into a more convenient form for charging the blast furnaces (Fig. 28, 5). The furnaces are charged automatically. Skips carry loads of coke, sintered ore and limestone to the top of the furnace and tip them in. The purpose of the limestone is to remove impurities, which float to the surface and form slag. When the furnace is tapped, the slag is run into pits where it is broken up for use as road-making material. The molten iron is cast by removing a clay plug at the base of the furnace. The white-hot iron runs out into ladles and the taphole is plugged again by remote control. From these ladles, the iron is poured into two giant mixers holding 1,200 tons of hot metal each.

So far, only iron and not steel has been made. It is in the steel plant, numbered 6 on Fig. 28, that the molten iron is manufactured into steel in converters, each charge of 130 tons taking 45 minutes. From the steel plant, ingots up to 30 tons go to the slabbing mill (Fig. 28, 7) where they are reduced to slabs of about 6 inches in thickness. In the hot strip mill (Fig. 28, 8), the hot slabs pass through a series of rollers until they are reduced to a long strip from about $\frac{1}{20}''$ to $\frac{3}{8}''$ thick. Fig. 30 shows the last

30. Photograph of stand in hot strip mill

stage in the hot mill, a stand of rollers through which the strip is passing at 37 miles per hour, to be subsequently cooled by water. No operators can be seen in the photograph because this mill is completely automated and controlled by a computer which is "fed" with the requirements for each customer's order.

Alongside the stand in Fig. 30 is a further set of rollers ready to replace the ones being used. This change of rollers is carried out once during every shift with the aid of the gantry crane. The hot mill, which is over half a milc long, is supervised by a crew of six men. The hot rolled steel can be sold in the form of sheets or coils for such uses as railway wagons and decks of ships. Any steel which requires a finished surface of a high standard such as a refrigerator or a motor car body, passes through the cold mill where it is further reduced to thicknesses ranging from about $\frac{1}{10}''$ to $\frac{1}{100}''$.

In the older type of rolling mill, steel sheets were passed by hand through rollers from side to side until they were compressed to the thickness required. The modern mills, in which the slab of steel is passed through sets of rollers without a break until it becomes a long strip of metal, are known as continuous strip mills. There are five works of this kind in the United Kingdom, of which four are in Wales. The first was built at Ebbw Vale in 1938. In 1951 the Margam Abbey plant at Port Talbot was completed on what had been an empty stretch of coastal sand. Another was built at Shotton in Flintshire on sands bordering the Dee estuary. Spencer works is the most recent of this type. Since these enormous plants were built, most of the older and smaller rolling mills in South Wales have closed down. South Wales has had a long tradition of rolling steel and tinplate and still specializes in the making of sheet steel. Steel now employs more workers than any other industry in Wales and South Wales produces more steel

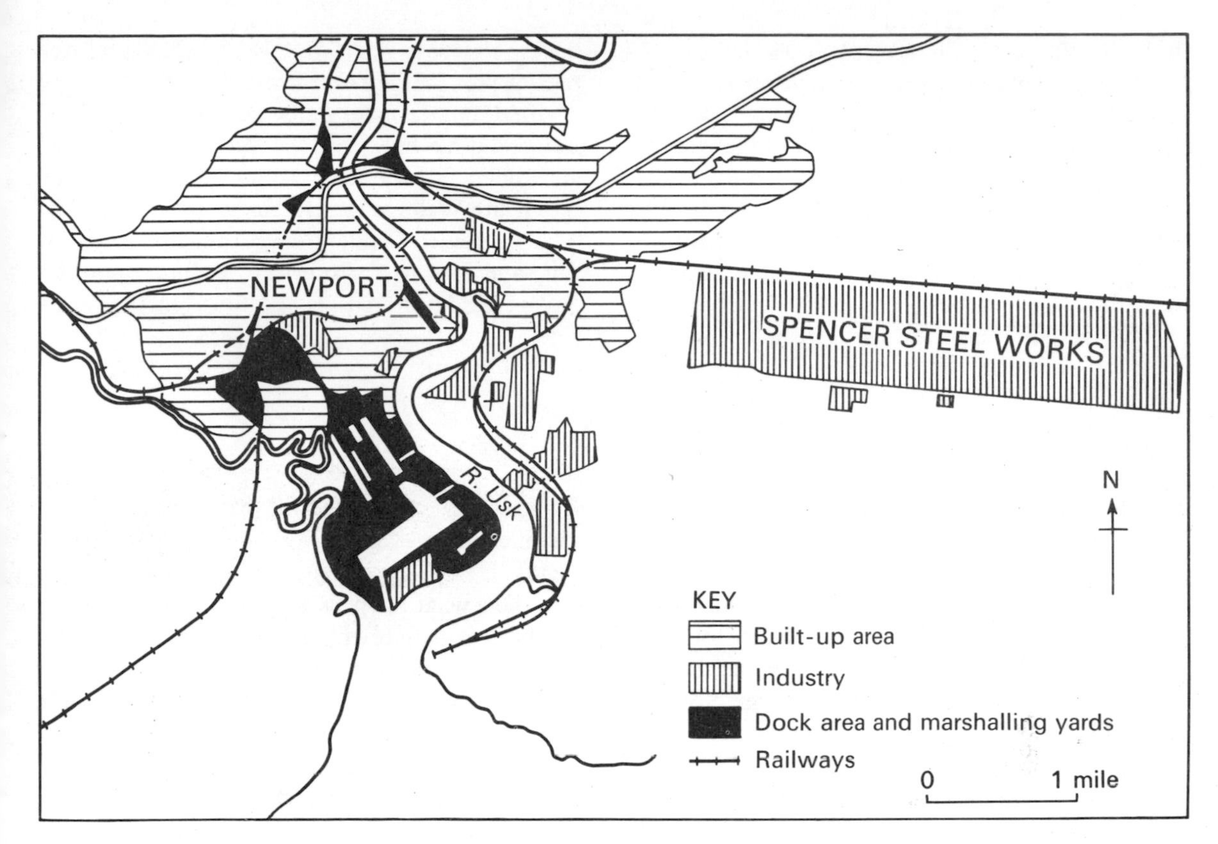

31. Map of Newport

than any other region in the United Kingdom.

From what has been said above, the main requirements of a modern steel works can be summarized.

1. A large clear site is needed since a steel works the size of Spencer works is equal in area to a town.

2. Most of the iron ore needed to make steel comes from abroad. In order to reduce to a minimum the costs of transport, the works should be as near as possible to the sea.

3. The works must be as near as possible to a coalfield to reduce the cost of the enormous amounts of coking coal needed. All the large steel works in Wales are on or near the coalfields.

4. Large supplies of water must be available either from the sea or from rivers. At Port Talbot the bulk of the water is pumped from the docks.

5. A modern steel works employs thousands of men. It must be possible to attract large numbers of workers, many of whom must be highly skilled. South Wales has a long tradition in the manufacture of iron and steel and hence it is possible to obtain the skilled men. The use of computers in steel manufacture also creates a need for new kinds of technical training.

6. Markets for the steel must either be close at hand or it must be possible to transport the steel cheaply to distant markets. There is a large demand for South Wales' steel in South Wales itself, especially in the tinplate industry. Large quantities are also used in the motor-car factories in the English Midlands which are connected to South Wales by a good railway route and by road. South Wales can export cheaply because most of the steel works are near the sea.

Not all these conditions hold good in the case of Ebbw Vale, where the works are situated in a deep valley eighteen miles from the sea. These works were built in 1938 when demands for steel were much lower and there was much unemployment in the Ebbw Vale area. The steel plant was built with Government help to provide employment.

Spencer works adjoins Newport, a town which is rapidly expanding in the direction of the works. There are some older metal industries in the town making steel, machinery and tubes, mainly oil pipelines. As the map (Fig. 31) shows, the factories in Newport are much smaller in size than the Spencer works.

Some shipbuilding has also grown up at the docks. Newport was formerly the chief coal exporting port for the eastern part of the South Wales coalfield, but the main traffic now is the import of iron ore. It has links with the Midlands of England and commands the main railway and road routes into South Wales at the crossing points of the Usk. The journey to London is just over two hours by rail, and the M 4 motorway leading from the Severn Bridge by-passes the town on the north side.

Exercises 1. Draw a sketch map of the main sections of a steelworks and strip mill (based on one of the three large plants in South Wales). Name the different parts of the works and show by arrows the movement of the raw material and products.

2. Draw a set of simple diagrams in sequence to show what happens in a blast furnace, a converter and a strip mill. Name the materials used and explain briefly the operations carried out. (Pamphlets of the type issued by steel firms may be useful in answering this question.)

3. Compare Spencer, Margam Abbey and Ebbw Vale works from the point of view of their sites and sources of materials.

4. Draw a sketch map of South Wales. Show the extent of the coalfield and the locations of the main iron and steel and tinplate works. Show the ports through which iron ore is imported.

5. Using Fig. 28 and the photograph (Fig. 29): (a) State the direction in which the camera was pointing when the photograph was taken. (b) What is the area of the steelworks to the nearest square mile? (c) Why were the stockyards built at the western and not at the eastern end of the steelworks?

6. Compare the photograph (Fig. 29) with the plan (Fig. 28). Identify on the photograph the cooling towers, blast furnace, the steel plant and the strip mills.

7. Describe what is happening in Fig. 30 and say where this fits into the succession of processes by which strip steel is manufactured.

8. What are the requirements of a modern steel plant and how do these influence the location of the steel industry in Wales?

10 | CARDIFF

There is usually a reason for a city being where it is. The old town and castle of Cardiff grew up where the route from east to west crossed the river Taff, near its upper tidal limit. The Taff estuary was in the most central position to deal with the export of coal from South Wales, and Cardiff became a great coal port. Docks were built on coastal flats bordering the Bristol Channel and the city expanded over the land between the Rhymney and Taff rivers.

One part of the city differs from another mainly according to the character, age and purpose of its buildings. The most important shopping centre contrasts with the dock area, and the older streets near the centre with those in the new suburbs.

The map (Fig. 32) and the photograph (Fig. 33) help to locate the distinctive zones of Cardiff. Some of these zones are seen in Fig. 33. The civic centre is in the foreground near the castle; beyond it is the commercial and shopping area with a greater density of tall buildings, and in the distance can be seen the docks and the steelworks. The beginning of a residential area appears to the left of the civic centre and the railway yards.

The civic centre, for which Cardiff is so widely known, is area 1 on Fig. 32. The buildings, surrounded by gardens, were laid out according to plan. Some, like the City Hall, are concerned with the government of the city; others, such as County Hall, with the County of Glamorgan. Other buildings, for example the National Museum of Wales, serve the whole country. Alongside the civic centre is the modern restoration of the Norman castle (Fig. 32, 2). Part of the former moat is now a car park (Fig. 33). The castle was situated where the main road crossed the Taff, and it controlled this crossing place from the earliest days. The flood plain of the river (Fig. 32, 3) is covered by parks. South of the road bridge, on artificially raised land, is Cardiff Arms Park, containing the rugby and cricket grounds.

In contrast to the spacious civic centre is the crowded commercial quarter, marked 4 on Fig. 32. A larger scale diagram of this area is shown in Fig. 34. Most people who come to Cardiff to shop know this part of the city best. Queen Street and St. Mary Street contain the large department stores. Away from these busy main streets shops give way to warehouses and offices, such as insurance and house agents. Many older buildings have been pulled down and the resulting spaces used as car parks. Although there are still a few streets of residential houses in the centre of Cardiff, most of the people who lived there have moved out to the suburbs. Some of the former large houses are now used as offices. Most of the commercial centre has its limits to the east and south along the main railway lines, but modern office blocks have recently been built to the east of Queen Street station.

Nearer to the General railway station is quite a different district. Here the streets are crowded with fruit and vegetable lorries, and nearby are small cafés used by the drivers. In Custom House Street the buildings are fruit warehouses and wholesale stores. This is the "Covent Garden" of Cardiff which grew up here because it is both near the docks and the railway station. A new market is being built at Leckwith on the outskirts of the city where traffic congestion is less liable to occur.

Leaving the central area and going south of the main railway into Bute Town (Fig. 32, 5), we enter a different world of docks and heavy industry. Bute Town is isolated not only by the railway, but also by Bute Docks to the east and the Glamorgan canal to the west. Before the railways and docks were built, parts of this area, for example Loudon Square, were a fashionable residential district. When Cardiff became a large port, an immigrant population of seamen settled there. Many of the present inhabitants are Muhammedans and among the buildings of Bute Town is a mosque. The men spend long periods at sea, followed by a few weeks

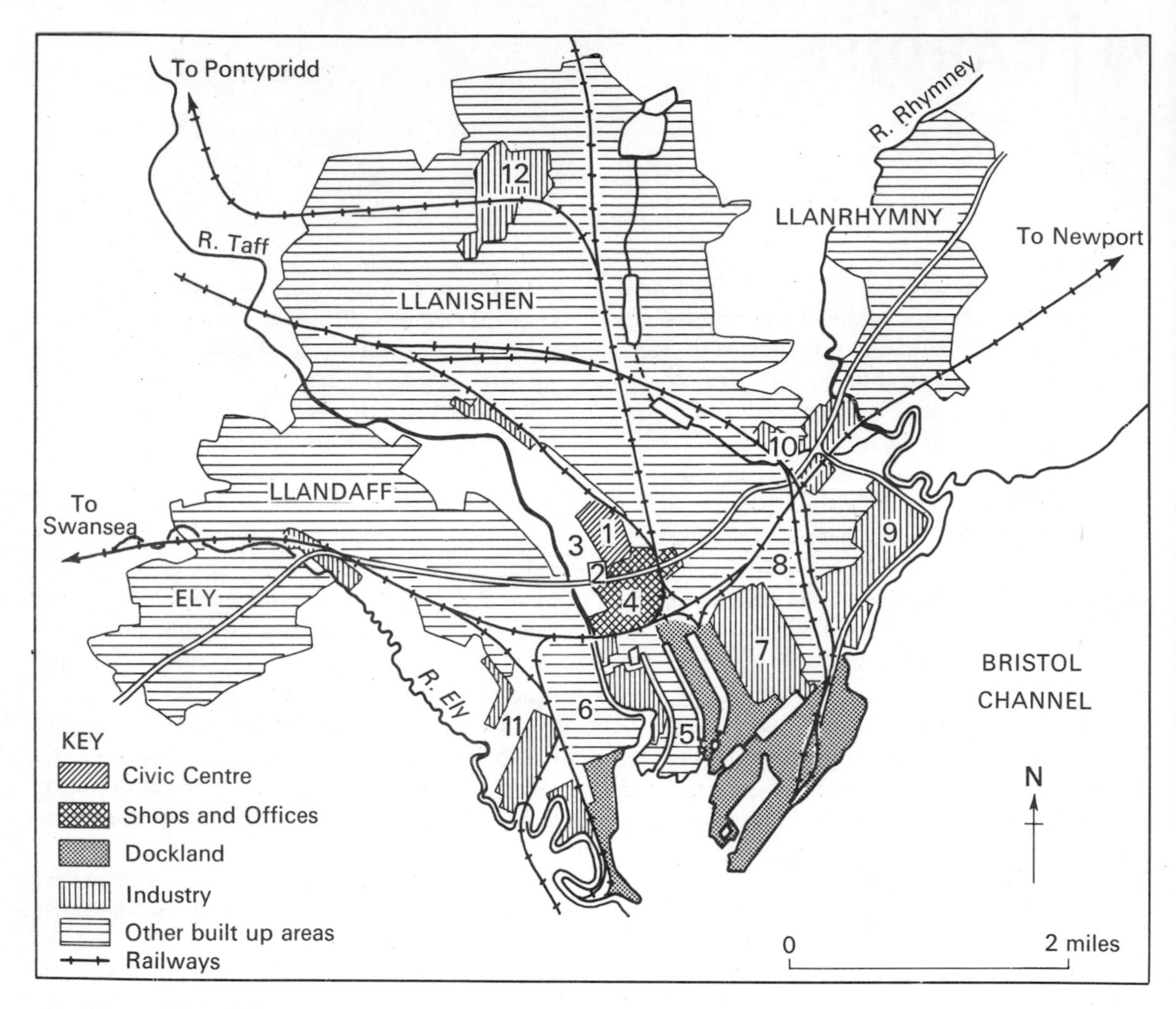

32. Map of Cardiff

or months at home in Cardiff. Because of their work, it is not convenient for them to move to the outskirts of the city as many people have done. The appearance of Bute Town has undergone a great change as tall blocks of flats have replaced the old streets of terraced houses.

Grangetown (Fig. 32, 6), where many dockers and industrial workers live, is also a self-contained area, bounded by railways to the west and the river Taff to the east. The grey stone houses are rather larger and not so old as those of Bute Town. Each district of Cardiff has its own shopping centre which, in Grangetown, is situated where several of its streets converge.

The lower Taff, flowing past Grangetown, is a tidal estuary where the first port of Cardiff grew up. Heavy industries, for example foundries, now border the river on the east side. When Cardiff became a great coal port in the last century, docks were excavated in the soft mud flats along the edge of the Bristol Channel. The most familiar sight in the docks used to be the coal hoists, which tipped loads of coal into the holds of ships. These can no longer be seen since Cardiff has ceased to export coal. The docks have large areas of railway yards, partly as a result of the use made of rail transport by the coal trade. Today road transport is of great importance, and a new road, "Rover Way",

33. Air photograph of Cardiff

gives access to the docks from the east side.

Dockside location is an advantage to heavy industries using large quantities of heavy raw materials. Alongside the docks, for this reason, are the Guest Keen and Nettlefolds iron and steel works (Fig. 32, 7).

In the shadow of the steel works are the terraced houses of Splott (Fig. 32, 8), which is hemmed in by railways on two sides. This district originally housed employees of the steel works, where many of its inhabitants still work. The newer housing estates of Splott are on the east side of the railway. Here there is also a different type of factory, the Rover car factory (Fig. 32, 9), built on the site of the former city airport, which has moved to Rhoose some miles west of Cardiff.

Other factories in Cardiff are mostly grouped in three industrial estates. The Colchester Avenue estate (Fig. 32, 10) is on low-lying land near the river Rhymney and flanks the main road entering Cardiff from the east. The Leckwith Moor estate (Fig. 32, 11) is on artificially raised land bordering the river Ely. The Birch Grove estate (Fig. 32, 12) is in a northern suburb. These industrial estates tend to use land in little demand for housing and are dependent on electrical power and road transport. The factories make a great variety of products, some of which have a link with the traditions of

Cardiff as a port. A typical list of manufactures includes pulley blocks (those used on the St. Lawrence seaway of Canada were made here), mechanical loaders and diggers, metal window frames, caravans, electric light bulbs and cigars. The majority of factories on these estates can be classed as light industry.

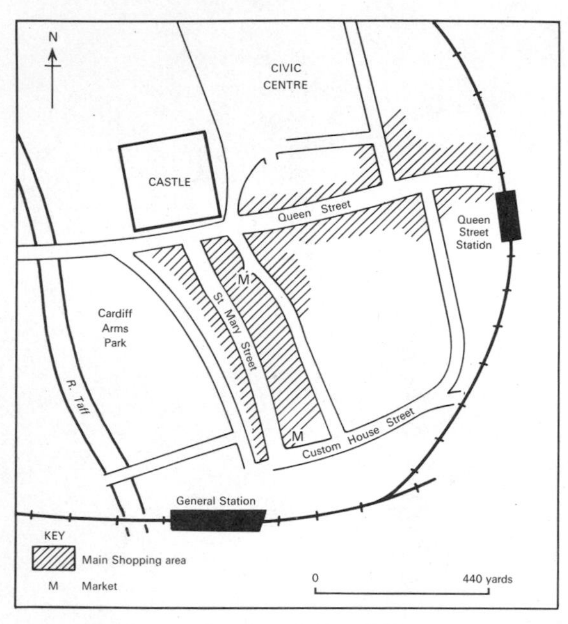

34. Diagram of central Cardiff

Most of the remaining area of Cardiff consists of residential suburbs. Those which are nearer to the centre are usually older, for example Roath and Canton. The latter houses many workers from British Railways' diesel-engine depot. The newer estates are found in the outer parts of the city in Ely, Llanishen, Rhymny and Llanrhymny. Among them are city council estates, occupied by people who have moved out from the centre, and private estates. Similar estates have also grown up outside the actual city boundary.

One of the characteristics of a city is movement. People must travel to work from one part of Cardiff to another and from a wide area around the city. In the morning rush hour they go to work in the docks and factories, in the shops and offices, and to schools. Shoppers and sightseers come into the centre of Cardiff. Goods are delivered to the docks and to shops and warehouses and to the markets. Cars, buses and trains carry thousands of people into and across the city each day. All this has to take place in a network of streets which has grown up in the past and was not designed for the present age of the motor-car.

Another characteristic of the city is growth. Cardiff, like other cities, has overflowed its earlier limits. It grows upwards in the form of flats and high office blocks. It grows outwards, as suburbs spread into the surrounding countryside. The built-up area around Cardiff has reached the foot of Caerphilly mountain to the north.

The city is linked to its hinterland, or the larger area which it serves. People come into Cardiff to work from many of the central valleys of the coalfield. They come to shop from most areas of the coalfield and from the coastal lowlands. The evening papers printed in Cardiff circulate throughout South Wales and a daily paper over most of Wales. Goods are delivered from Cardiff warehouses and stores to the eastern and central parts of the coalfield. Goods shipped through the docks come from a still wider area, reaching as far as the English Midlands. All these movements depend on road and rail transport, using routes which converge on Cardiff.

In 1955 Cardiff, the largest city in Wales, became the capital. Not all the national buildings of Wales are in Cardiff, however. The reasons for this, and the reasons why Wales had for so many years no official capital, are questions considered in the final chapter.

Exercises 1. Imagine you are a driver of a lorry carrying one of the following: steel bars, Rover car parts, petrol, foundry-cast windows, beer, cigars, or fruit. In each case, say to which part of Cardiff you are going to or coming from. If a street map of Cardiff is available, show your route.

2. Draw a sketch map of Cardiff showing the shopping centre, the civic centre, the dock area, the industrial estates. Write a brief note on each of these districts.

3. Compare the map (Fig. 32) with the photograph (Fig. 33). In which direction was the camera looking? Find on the photograph the civic centre, the shopping centre, Guest Keen and Nettlefolds steelworks and the docks.

4. Find evidence from Fig. 33 for the following statements: (a) Cardiff imports oil, (b) heavy industry is located near the docks, (c) part of the city was planned, (d) railways are important dividing lines in the city.

5. Why is Cardiff so much larger than Abergavenny? What features, if any, have they in in common?

6. With the aid of Fig. 44 and an atlas, draw a sketch map showing the coast of South Wales from Swansea to the Severn Tunnel. Show the position of Cardiff by a small circle and draw the main rail and road routes which pass through it or converge on it.

7. Obtain a copy of the official handbook and a large-scale Ordnance Survey map or street map of a large town or city you know well. Then write one or two sentences about each of the following: (a) the main industries, (b) the residential areas, (c) the chief shopping centre, (d) the civic buildings, (e) the population.

8. Answer the following questions about the city or town in which you live or which is nearest to your home: (a) What kind of shops and chain stores are found in the shopping centre; are the shops in some parts of the centre different in character from those in others? (b) What different kinds of housing estates or residential areas occur in the town; where are the poorer and where are the wealthier types of house? (c) Is there a traffic problem?; if so, give details and state remedies which have been proposed to solve the problem.

11 | SWANSEA

The visitor to Swansea who climbs Kilvey Hill is rewarded with a wide view, not only of the town but also of a large part of Glamorganshire. On the northern horizon stand the isolated masses of the Brecon Beacons range, lying beyond the plateaux of the coalfield. To the south is the full sweep of Swansea Bay from Port Talbot to the Gower Peninsula and, on the far side of the Bristol Channel, the coast of Somerset and Devon. Within closer range (as shown on Map D), the valley of the Tawe leads towards Swansea, narrowing as it passes between Kilvey Hill (Map Ref. 671940) and Town Hill (Map Ref. 641939).

The narrow coastal plain at the foot of Kilvey Hill broadens westward where the main residential area of Swansea has expanded beyond the Tawe. A continuous urban belt also occupies the western side of the Tawe valley as far as Morriston (see Map D, Map Ref. 665978). Housing estates have spread over a large part of Town Hill with the result that some of the Swansea streets are the steepest to be found in any town in Britain. The

35. Air photograph of Swansea Docks

town centre, near the western side of the Tawe estuary, was destroyed during the Second World War. Part of what appears on Map D as open space has been rebuilt since this map was published.

On the narrow coastal plain at the southern foot of Kilvey Hill is the district of Port·Tennant and beyond it are the modern docks shown in the photograph (Fig. 35) and in the sketch map (Fig. 36). The Prince of Wales', King's and Queen's docks were excavated in the sandy foreshore east of the Tawe estuary between 1890 and 1920. The first port of Swansea was in the sheltered estuary of the river, which at that time swung round to the west of the feature named "The Island" on Map D, Map Ref. 660934. In the early days of industry in the Swansea valley, metal ores were unloaded along that stretch of the estuary.

Probably the most prominent features in the modern dock shown in the photograph are the coal hoists. The reorganization of the South Wales ports has resulted in most of the coal exported from South Wales being directed through Swansea, and coal hoists have been transferred there from other South Wales ports. The amount of coal exported is much less than before 1939, but anthracite forms a substantial part of it. Apart from coal, many of the exports come from the western half of the South Wales coalfield, and high on the list of these are metal manufactures and tinplate. Steel sheets are a valuable export, and being a clean, compact form of heavy bottom cargo, are preferred by the liner cargo ships. Among a long list of other exports are the manufactures of the Fforest Fach Industrial estate and the factories in the Swansea area. An important item is dry-cleaning machinery and a minor one Corgi toys.

Oil tankers berth at the Queen's Dock and nearby are oil storage tanks. Swansea was the first large

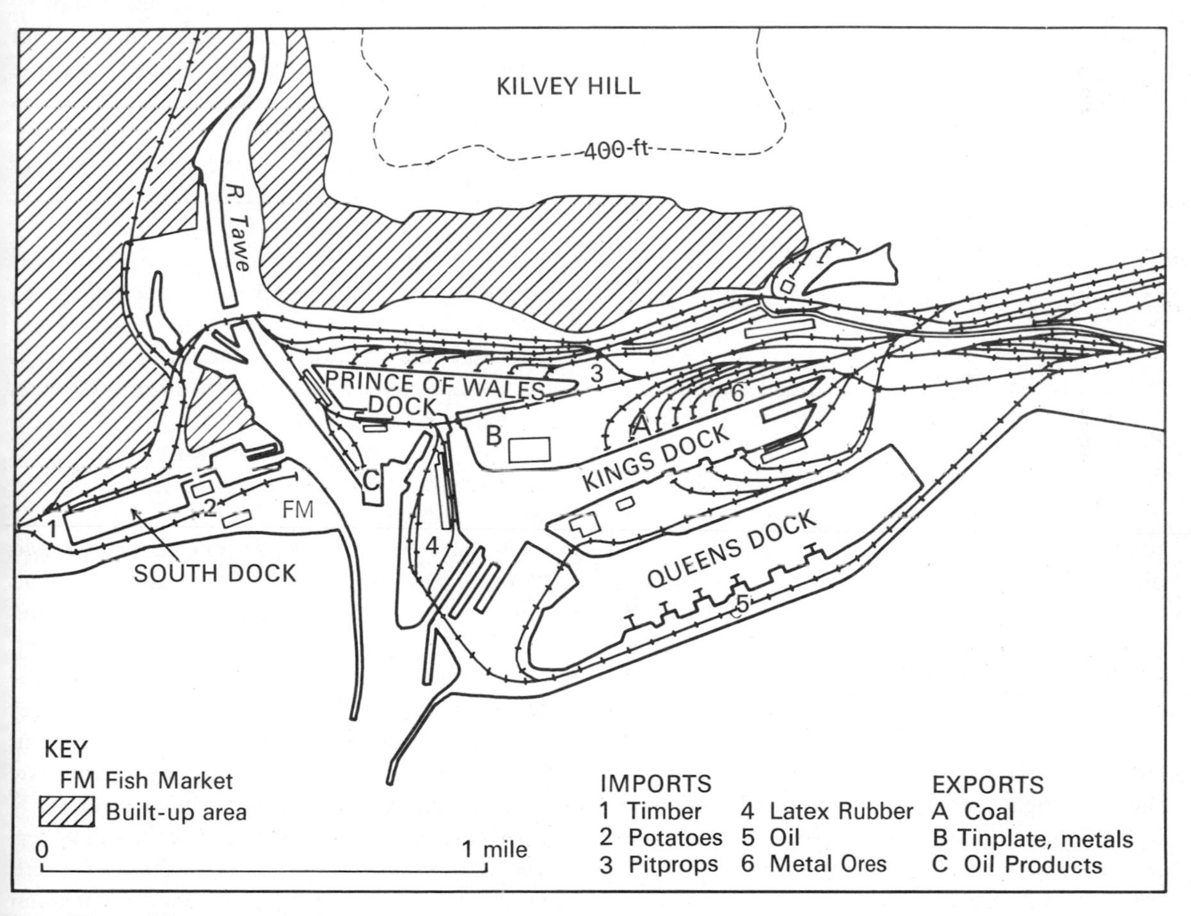

36. Map of Swansea Docks

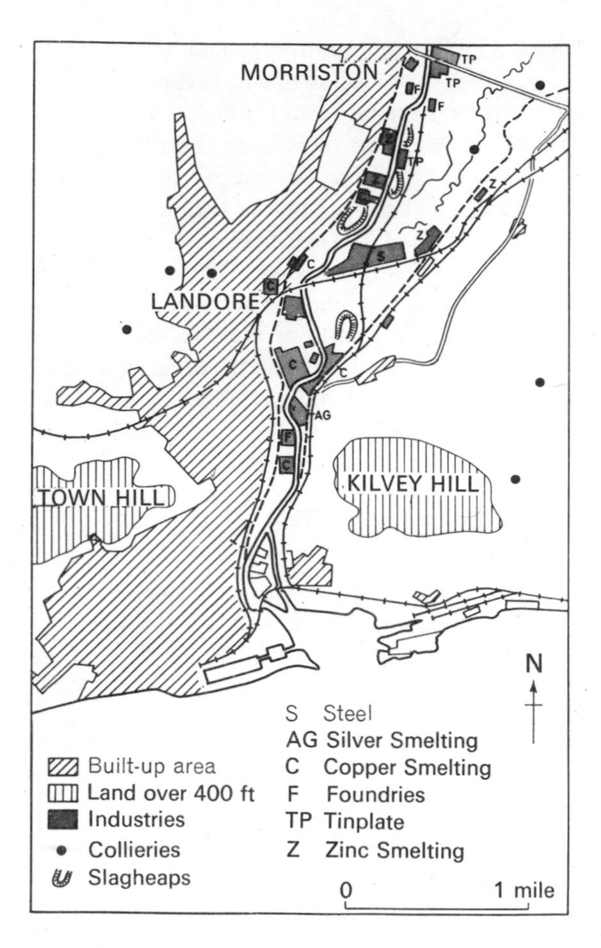

37. Swansea and its industries in 1884

oil port in South Wales, but the oil terminals in Milford Haven have recently reduced it to second place. The crude oil is imported and piped a short distance to Llandarcy refinery, which is also connected to Milford Haven by a 68-mile long pipeline. Refined oils, such as petrol, diesel oil and lubricating oil, are also distributed from the port of Swansea to different parts of the British Isles. Oil is not the most valuable import, however. First comes nickel ore, which has already been partly treated in Canada where most of it is mined. This goes to the famous Mond nickel works at Clydach. Other imports are rubber latex, unloaded at one part of King's dock (see Fig. 36), and timber pit props, in small quantities only since many of these are now made of steel. At the old South Dock, Irish potatoes are unloaded, including a large kind specially favoured for making chipped potatoes. Fish arc brought into the same part of the dock, for Swansea is also a fishing port.

The fortunes of the port are linked with the industries of the Swansea area and their fluctuating history. For centuries, sailing ships brought copper and lead ores from west and north Wales to the estuaries of the Llwchwr, Tawe and Neath, where the coalfield flanks the sea. Tin ore also came from Cornwall. All these ores were smelted with the local coal. When large new deposits were found in America, Africa and Australia, these raw materials were still transported to the Swansea area where the smelting industries had become established. Most of the smelting works were in the lower Tawe valley and the largest concentration was in that part of the valley shown on Map D. The two railway lines running down either side of the valley formed the approximate boundaries of the industrial area. Between them was an almost continuous belt of mills and slag heaps. The furnaces smelted copper, tin, lead and zinc ores and emitted chemical fumes which destroyed much of the vegetation on the valley sides. Fig. 37 shows the works existing in this part of Tawe valley in 1884.

Metal smelting began to decline when overseas countries which supplied the ores began to smelt their own metals, but it did not disappear immediately because the tinplate industry was expanding during the latter half of the nineteenth century. Llanelli became the main centre of the tinplate industry, but many works also grew up in the Swansea area. The famous Siemens steelworks, where the Siemens-Martin furnace was developed, was also located in the Tawe valley (see Fig. 37) and its production was closely associated with the tinplate industry.

At that time the population of Britain had grown to such an extent that more and more food had to be imported from America and Australia. Canning helped to solve the problem of preserving such foods as fruit and meat. American meat producers had already begun to make large quantities of tinned "bully beef", used during the American Civil War to feed armies in the field. The raw material needed for canning was tinplate and nearly all the tinplate in the world came from South Wales, where the process of coating thin sheets of steel with a very fine layer of tin to protect them from corrosion had been perfected.

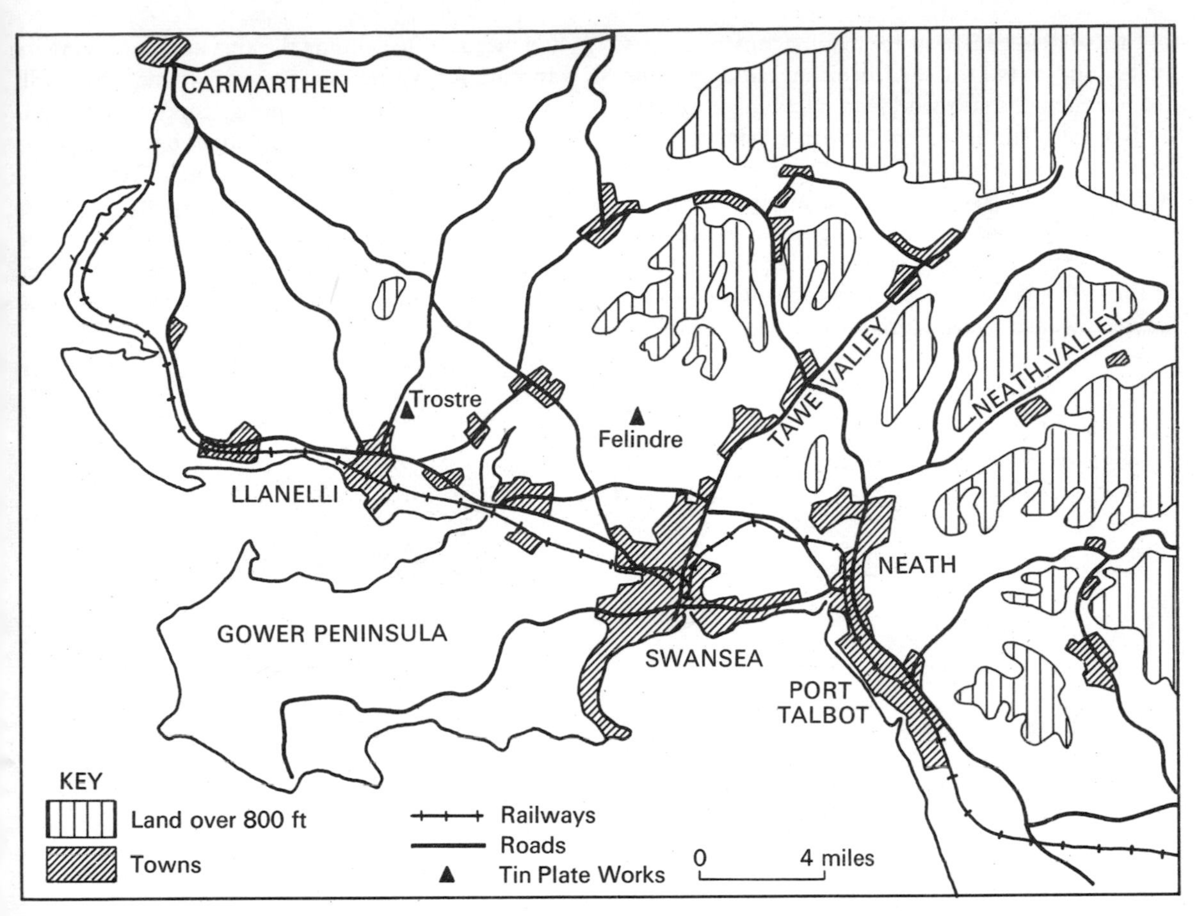

38. Map of the Swansea area

Soon, the United States began to manufacture their own tinplate and other countries followed, so that towns like Llanelli, which were largely dependent on the industry, began to suffer temporary loss of trade. Tinplate manufacturing has continued to expand, however, in spite of the invention of other materials like plastics. A glance at the shelves of any supermarket shows the extent to which tins are still used, and beer and mineral waters are now canned as well as bottled. The entire British production of tinplate sheets, amounting to roughly one million tons, is manufactured in South Wales. The majority of factories which make the actual cans and metal boxes are, however, located in England.

The organization of the tinplate industry is now completely different from what it has been in the past. Formerly there were over thirty tinplate works and rolling mills in south-west Wales. Workers passed the steel sheets through rollers from side to side, manipulating them by hand. The sheets were then dipped in vats of molten tin. Today manufacture is concentrated in three large works at Ebbw Vale, Trostre and Felindre; the last two are situated in the south-west of the coalfield (see Fig. 38) and receive coils of strip steel from the Margam Abbey steelworks at Port Talbot. The coils are further rolled in the Trostre and Felindre strip mills and coated with tin by a continuous process.

Today the face of the lower Swansea valley has completely changed. It has no tinplate or steel industry. Workers in these industries travel to the two large mills which have been mentioned. No copper is now smelted at Swansea and the last relic of the copper industry is the manufacture of copper plates and vats still carried on at Landore. The former Siemens steelworks produces ingot moulds and furnace lining bricks for other steelworks in South Wales (Map Ref. 670961). In the old copper works, packing cases and ventilation equipment are made and some of the industrial buildings are now depots for transport and steel-erecting firms. Much of the valley is derelict. An organization called the Lower Swansea Valley Project is at work examining means of reclaiming the area and improving its appearance. Varieties of grass are found which will colonize slag tips containing copper and lead. Trees are planted and eventually the valley may be covered with playing fields and new factories instead of industrial ruins. At present, however, Swansea, unlike Cardiff, is not short of space for expansion within its own boundaries, and other areas attract development more easily than this derelict land.

Exercises Use Map D in answering questions 1 to 7.

1. (a) Draw a north to south section across Kilvey Hill from Pentre-chwyth to the docks, on a vertical scale of 1 inch to 500 feet; (b) explain the shape of the section, bearing in mind that the hill is of Pennant Sandstone and lies on the southern edge of the coalfield.
2. Contrast the view you would see looking north from Kilvey Hill with that looking south.
3. How does the urban and industrial development of the area west of the railway from Swansea to Morriston differ from that to the east?
4. Draw a sketch map of Swansea, dividing it into: (a) the dock area, (b) the industrial districts, (c) the town centre, (d) residential and suburban districts. Show land over 400 feet.
5. (a) Describe the physical and human changes in the character of the Tawe valley from north to south; (b) draw two simple annotated cross-sections illustrating the differences between the northern and southern parts of the Tawe valley.
6. Compare Fig. 37 with Map D and outline the differences between Swansea today and Swansea in 1884.
7. Draw a sketch-plan of Swansea docks, adding notes on the purpose of some of the features shown on the map.
8. Write an account, illustrated by a sketch map, of the industries within approximately 10 miles of Swansea.

12 | A GENERAL VIEW OF WALES

Relief

As Fig. 39 shows, much of Wales is hilly or mountainous. Only in Anglesey, south-west Pembrokeshire and the southern part of Glamorgan are there extensive stretches of lower land. Even there, most of the land below 400 feet in height is not flat coastal plain but low plateau, dissected by valleys, and often ending at the sea in cliffs, like those which appear on the photograph of Milford Haven (Fig. 46).

The highest parts of Wales, over 2,000 feet in Fig. 39, are in two main areas. The first is Gwynedd, or north-west Wales, where the mountains, from Snowdon to Cader Idris, are rugged, glaciated and mostly made of hard slates and volcanic rocks. The second area contains the sandstone scarps of the Brecon Beacons range and the Black Mountains of South Wales. The remaining upland areas, over 1,000 feet in Fig. 39, are plateaux with fairly even surfaces, dissected by steep-sided valleys.

Valleys penetrate into the heart of Wales. Three of the longest of them, the Dee, the Severn and the Wye, lead into England.

Climate

Mist descends on the hillsides, followed by drizzle and rain, then showers and cloud shadows move rapidly over the hills; this is a common

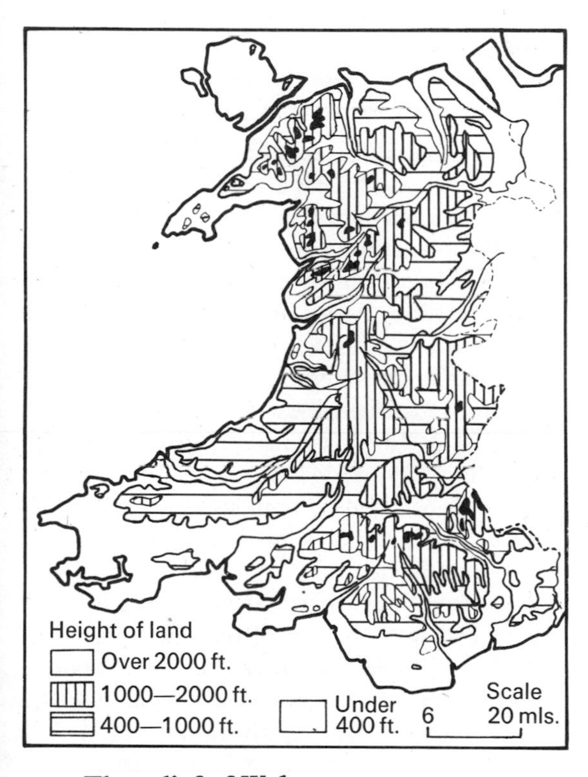

39. The relief of Wales

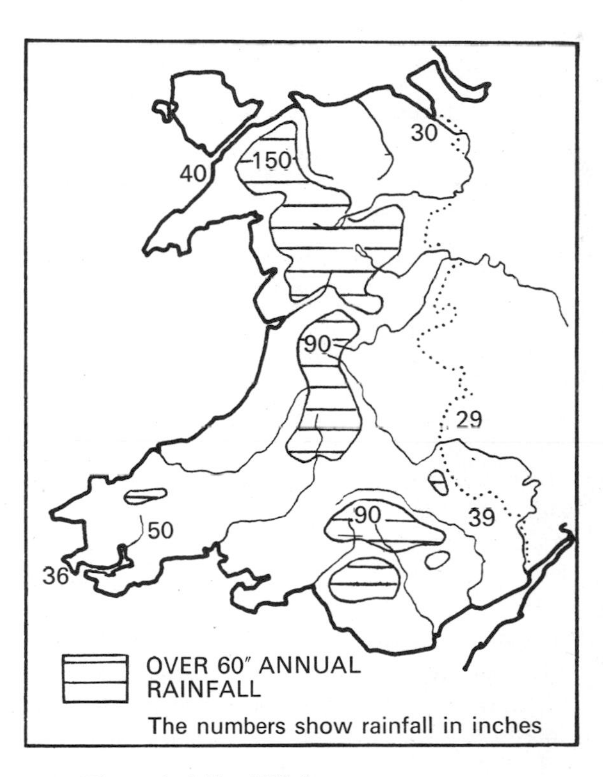

40. The rainfall of Wales

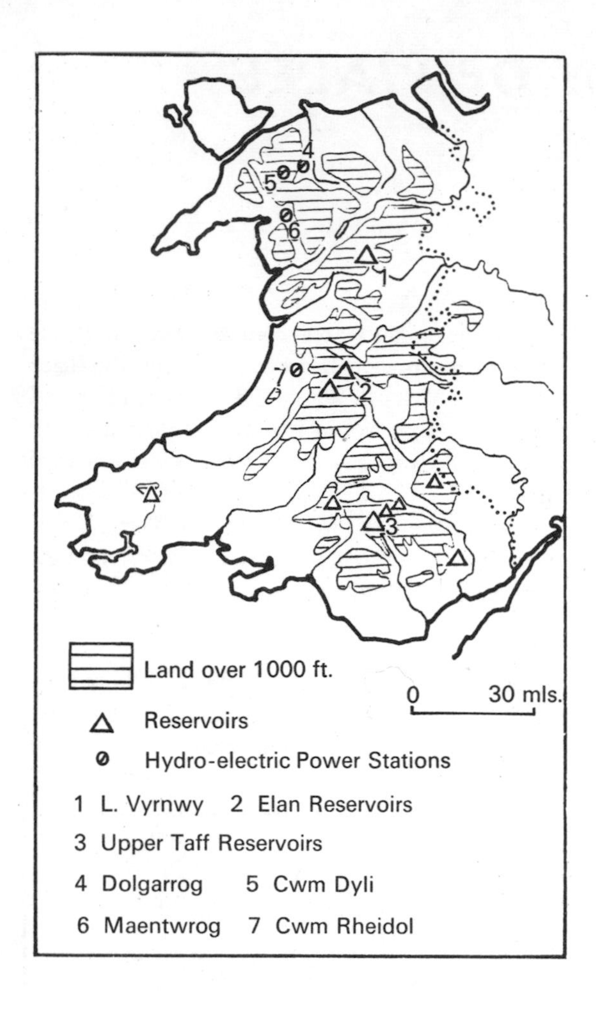

41. Wales, water supply

sequence of weather in Wales and marks the passage of a depression or rain belt. Wales owes its plentiful rainfall to the depressions which approach or cross the country from the Atlantic Ocean and the Irish Sea. The mountains force the air to ascend and cause an increase in rainfall. Thus, while coastal areas have an annual rainfall of 40 inches or less, the moorlands have 60 inches and high mountains over 80 inches (Fig. 40).

A comparison of Fig. 40 with Fig. 39 shows a close connection between relief and rainfall. The westerly movement of rain belts across the country, combined with the influence of the mountains, results in a greater rainfall in the western half of Wales than on the eastern border adjoining England. The eastern margins include some rain

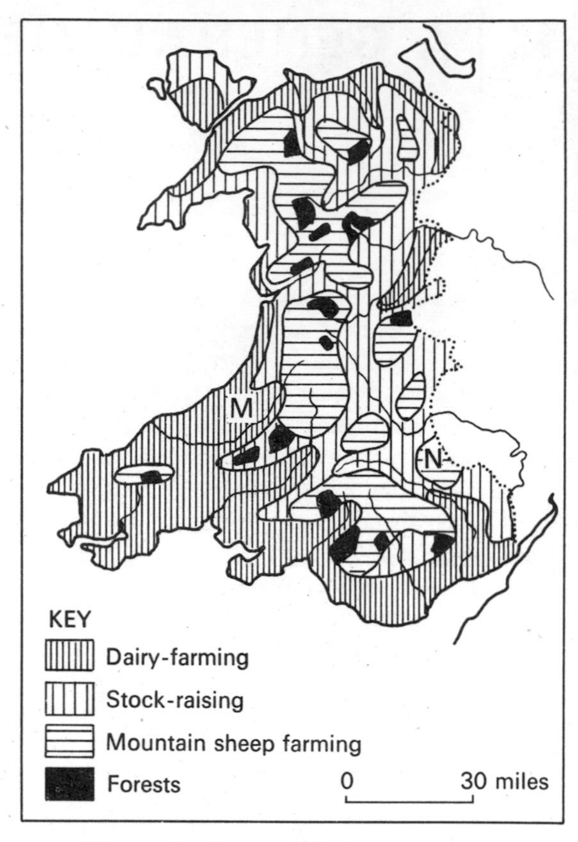

42. The agriculture of Wales

shadow areas, which would have a greater rainfall if the mountains did not attract so much.

The large amounts of water provided by rainfall over the uplands form one of the natural resources of Wales for which there is much demand from towns and industry in England and South Wales. As the examples of the Rheidol and Elan dams show in Chapter 3, the water is used both for generating electricity and for piped water supply. Fig. 41 shows the location of the main reservoirs in Wales. Water from Central Wales goes to Liverpool and the English Midlands. The industrial area of South Wales obtains its water from the Brecon Beacons and Black Mountains.

No part of Wales is more than 50 miles from the sea, and the western part of the country owes its mild winters to sea influences. The eastern border of Wales, being further from the sea and in the lee of the mountains, has slightly colder winters. The mountains have different climatic conditions from the lowlands, and experience lower temperatures and more winter snow.

Farming and Forestry

The uplands of Wales are difficult for any kind of farming other than stock raising; in those areas sheep farming predominates (see Fig. 42). Thus the high moorlands and mountains are ringed around with sheep farms, of which Neuadd, described in Chapter 5, has been taken as an example. While sheep are reared in most parts of Wales, the highlands specialize in sheep farming. From these same districts people have migrated, leaving behind deserted farms many of which have been amalgamated with others to make large sheep ranches. Forests have also been planted in upland areas.

There are nearly six million sheep in Wales and the number has been increasing in spite of losses due to severe winters in 1947 and 1963. One of the difficulties is the winter feeding of animals, and hay has to be brought into the sheep-farming areas. Upland farms are given subsidies by the government and the growing of more winter feeding stuffs is encouraged.

About two-thirds of Welsh farms produce milk for sale. The mild and moist climate of the western part of Wales favours the growth of grass. These were formerly areas of mixed farming, but when creameries and milk factories were built, many farmers took up dairy farming. Maes Mynach (chapter 4) is an example of such a farm. Fig. 42 shows the location of the main dairy-farming districts. From the west, the milk tanker lorries follow the main roads to London, the Midlands, and the industrial towns of South Wales.

The valleys of the Severn, Wye and Usk, opening eastwards towards England, contain farms which carry on stock raising, the rearing of store cattle and sheep for fattening in England. In these areas, which are drier than the western part of Wales, barley, root crops and kale are grown to feed cattle. The Welsh border, therefore, makes a very important contribution to meat production in Britain, and the trade is carried on through the market towns, such as Builth, Brecon and Abergavenny (chapter 6).

The movement of stock from Wales into England goes back to the days of the drovers, who gathered cattle from the western counties and moved them by regular routes to the border, stopping overnight at inns, some of them still known as the Drovers' Arms. The transport of cattle and sheep into England now takes place by lorry and amounts to hundreds of thousands of stock annually.

Industry

Reference was made in chapter 6 to the existence of woollen mills in the market town of Abergavenny a century ago. Many other towns had a woollen industry at that time and there are still fifty mills in Wales, all small concerns. Newtown, in Montgomeryshire, became the chief woollen manufacturing town in the last century, but most of the surviving mills are found in the Teifi valley and in south-west Wales.

The mining of metallic ores formerly took place in many parts of rural Wales. Two examples have been mentioned at Halkyn in Flintshire and Ystumtuen in Cardiganshire. Lead, silver, copper and zinc were the main metals mined and there was also a little gold mining in the Snowdon area. The ores were usually transported by sea to smelting

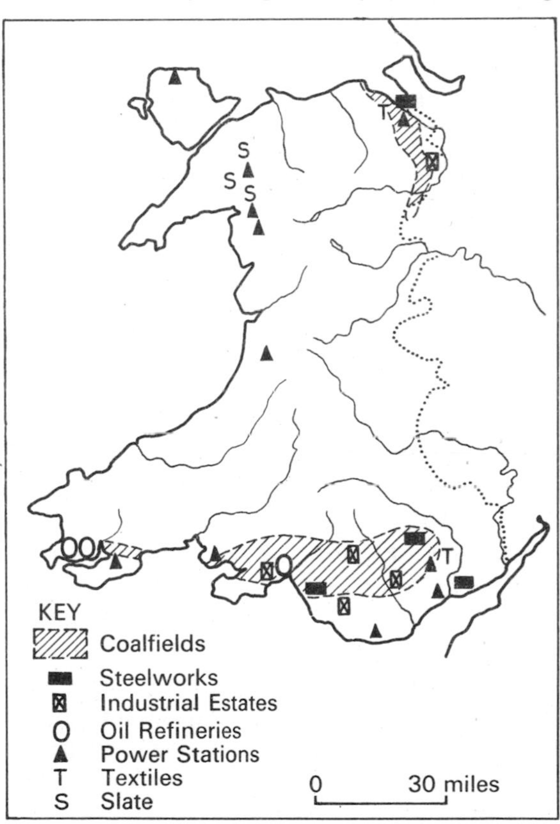

43. The industries of Wales

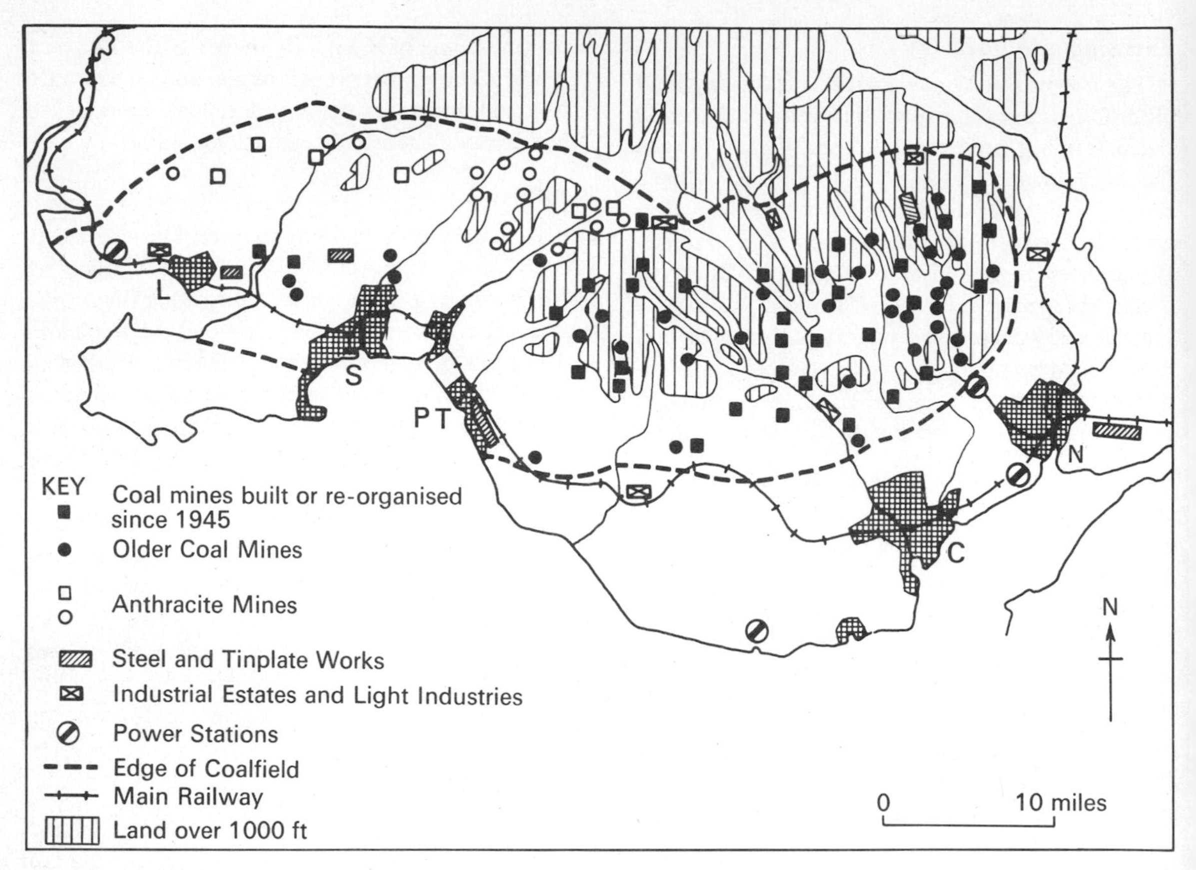

44. The industries of South Wales

works at Swansea. Many of the small mining industries had declined by the nineteenth century.

The industries of modern Wales are shown in Fig. 43, from which it can be seen that most Welsh industry is on or near the coalfields and that South Wales is more important than any other area. A more detailed map of the South Wales coalfield and the surrounding industrial areas is given in Fig. 44. Coal mining in South Wales has undergone many changes. Shallow mining started on the margins of the coalfield, especially on the northern side. As the coalfield is in the form of a basin (see Fig. 24), the productive coal seams lie at a greater depth in the centre of the coalfield. Deep mining extended into the central areas after 1850, and railways were built up all the valleys to carry the loaded trains downhill to the ports and to the main line to England. Mines were sunk at frequent intervals along the floors of the valleys which dissect the coalfield. Overlooking each large mine was a stone built town like New Tredegar (chapter 7). Immigrants flocked into these towns from the rural areas of Wales and the neighbouring counties of England.

Four types of coal were of economic value, house coal and coking coal in the eastern part of the coalfield, steam coal in the centre, and anthracite in the west (Fig. 44). The greatest boom in coal mining arose from the demand for steam coal by ships and railways. When electricity and oil replaced coal, the demand for steam coal declined. The Rhondda valleys produced steam coal and along them densely populated mining towns merged into one continuous belt of houses and mines. The photograph (Fig. 45) shows the results in one densely populated part of the Borough of Rhondda. The coal mine visible in the photograph is no longer active.

The anthracite mining areas had a different course of development. Anthracite could be used for smelting ores and was also in demand by in-

45. Air photograph of the Rhondda Valley

dustries and cities requiring smokeless fuel. Until recently the anthracite mines were confined to the northern edge of the coalfield, as deep mining in the anthracite zone was difficult.

Since the Second World War the mining industry has been nationalized and, as Fig. 44 shows, many collieries have been reorganized and some new ones built. The new mines have been built in parts of the coalfield where the coal was previously too difficult to mine. Thus, new deep mines like Abernant (see chapter 7) have been sunk in the central part of the anthracite field and have met with many technical problems. Nantgarw (chapter 8) was sunk near the southern edge of the coalfield where the seams dip steeply. Drift mines or levels, on a large scale, exploit some of the remaining coal in the older mining areas. Remote control of mechanical cutting at the coal face is being carried out at Cwm, near Ebbw Vale. The number of men employed in the mining industry has steadily declined, but production of coal has remained at a high level. Coal mining continues to be a substantial industry and is still vital to the nation. Electrical power stations consume over half the coal now produced

46. Air photograph of Milford Haven

in Britain and over a quarter is made into coke for blast furnaces. Industry and domestic users and the gas industry take most of what remains. Only a small proportion is exported, compared with nearly half in 1938.

Electric power stations supply householders and a large part of industry with power. The hydro-electric power stations already mentioned in this chapter supply some of this power, but the bulk of it comes from coal-fired generators and two nuclear power stations at Trawsfynnydd in Merionethshire and Wylfa in Anglesey. There is also an oil-fired power station at Pembroke (Fig. 47).

Heavy industry and, in particular, the iron and steel industries, have been closely associated with coal mining in South Wales. Merthyr Tydfil (see chapter 8) is typical of the iron-manufacturing towns which grew up on the northern margin of the coalfield where iron was mined. With the exception of Ebbw Vale, these towns have ceased to

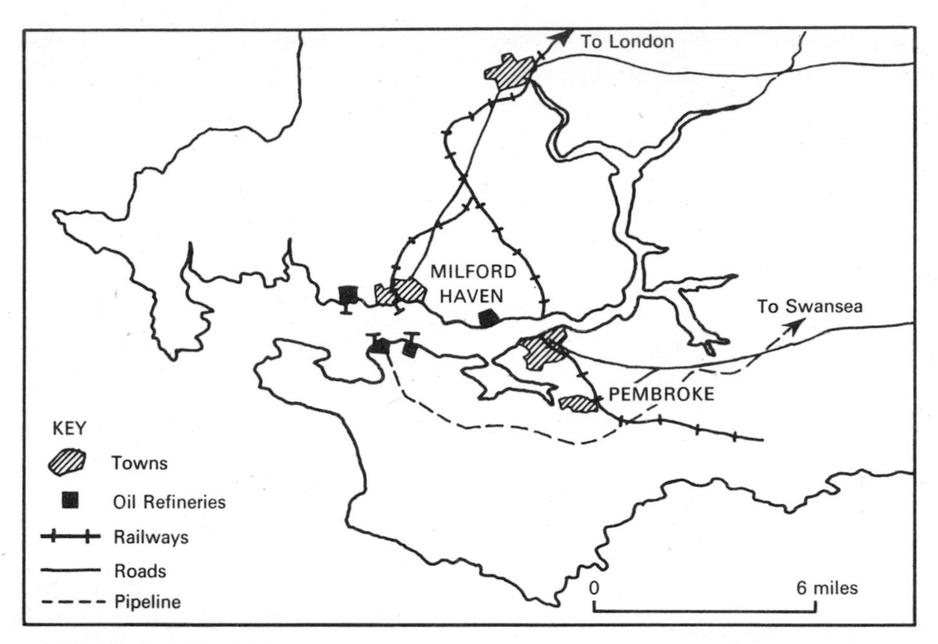

47. Map of Milford Haven

manufacture iron and steel, and larger iron and steel works on the coast have taken their place.

The vast modern plants which have arisen on the coastal plain at Margam and Newport (see chapter 9) follow the Welsh tradition of making sheet metal and tinplate, but mass-produce these materials on a greater scale than ever before. The steel industry in Wales now employs more workers than any other single industry. Too much reliance on heavy industries, however, led during the 1930's to unemployment which affected the two main industries, coal and steel. The numbers of unemployed in the mining areas amounted in some cases to between 60 and 90 % of the adult male population. The government took measures to remedy this situation by encouraging the kind of light industries which have prospered in the London area and the Midlands. The move to introduce a greater variety into Welsh industry led to the Treforest Industrial Estate in 1936. The study of this estate in chapter 8 shows the kinds of different products manufactured. Other industrial estates were built or adapted to wartime factories as Fig. 44 shows, for example at Hirwaun, Swansea and Bridgend. Subsequently, many other individual factory premises have been built throughout Wales in smaller towns and places threatened by unemployment, like Milford Haven and the slate-quarrying towns of North Wales. Some large factories like British Nylon Spinners at Pontypool and Hoover's at Merthyr Tydfil have had a considerable effect on the prosperity of those towns.

Oil refining, the rival of coal, has also gained a footing in South Wales. The first refinery was Llandarcy, near Swansea, but the main refineries are now around Milford Haven. This deep and sheltered ria, or drowned valley, is probably the best inlet on the west coast of England and Wales in which to berth large tankers of 100,000 tons or more (see Fig. 46 and Fig. 47). The Esso refinery and jetty, built in 1959, was followed by a B.P. oil terminal connected by pipeline to Llandarcy, and a Regent refinery and jetty. The Gulf Oil Company plans to build a third refinery. Neither Pembroke dockyard nor the fishing port of Milford were in a flourishing condition before the refineries were built. The port of Milford Haven now plans to build its own jetty for handling iron ore. The oil-fired power station at Pembroke has already been mentioned.

The other main industrial area of Wales, which is much less important than South Wales, is the North Wales coalfield area. Along the Dee estuary around Flint, coal mining is now of little significance, but the textile and steel works make this a busy coastal strip. Further south around Wrexham coal mining and steel making are still active and there is an industrial estate including a large

British Celanese factory. The proximity of the south Lancashire and north Cheshire industrial areas has helped the development of this part of Wales.

Transport

There are two routeways of outstanding importance in Wales (Fig. 48). They are the main road and rail route through Newport, Cardiff, Swansea and Llanelli in South Wales, and from Chester to Holyhead in North Wales. In each case the route follows the coastal lowland, which is wider in the south than in North Wales. The South Wales route carries Welsh manufactured products into England and to English ports through which many are exported. The Severn bridge and the extension of the M4 Motorway have improved the link with London. On the north side of the coalfield, the Heads of the Valleys Road is an additional route into South Wales and enables steel sheets and some coal to be carried to the M5 and the Midlands. The Chester to Holyhead route is more concerned with the passenger traffic to the North Wales seaside resorts and to Ireland via the packet station of Holyhead.

The narrow, congested roads in the mining valleys of South Wales are difficult for transport. A few of these roads come to dead ends, but seven of the valleys open northwards on to the Heads of the Valleys route, and a spectacular alpine road crosses Mynydd Beili-Glas into the Rhondda valley.

Population

Most people live where they are able to find a livelihood. Some live on farms which have been handed down from one generation to another, others in industrial areas to which their ancestors migrated or to which they themselves have moved. Fig. 49 shows the distribution of people in Wales. There is a central zone of highlands where the population is very sparse. Bordering the highlands are the farm settlements on the low plateaux and in the valleys. In strong contrast are the industrial areas and the North Wales coastal zone. 70 % of the population of Wales live in or around the South Wales coalfield, in an area containing about one and a half million people. About half the area of Wales has only 10 % of the population. The sparsest areas are also those which are losing people through migration.

Because so many of the people live on the borders of the country and because there are so few in the central districts of Wales, it has proved difficult to select one city as the capital. Thus, Cardiff was not officially proclaimed capital until 1955, after Swansea, Aberystwyth, Caernarvon and Machynlleth had all been considered. Cardiff has the advantage of being the largest town and a focal point for the most densely populated part of Wales, but it is remote from North Wales and outside the main area of Welsh speech. The National Eisteddfod continues to be held at a different place each year.

As Fig. 50 shows, the area where over half of the population are able to speak Welsh lies in the western half of South Wales, but covers a much bigger proportion of North Wales. In most of these districts the proportion is much higher than

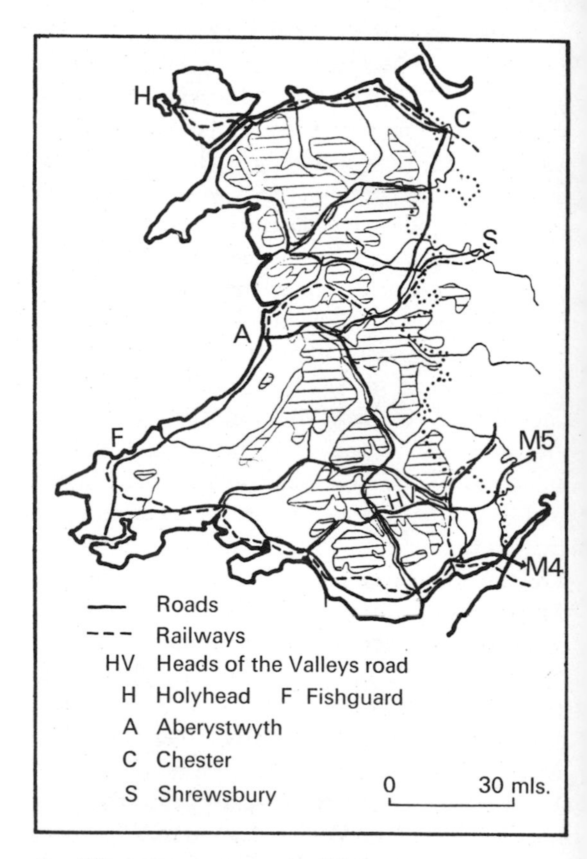

48. Transport routes in Wales

50 % and reaches 90 % or more in Anglesey, Llŷn and the uplands. Although Welsh is spoken in the slate-quarrying districts of North Wales and in the anthracite field of South Wales, most of the Welsh-speaking area is farming country. Here the Welsh tradition is based on family links, poetry, song and the Bible.

In the eighth century Offa, the king of Mercia, built a dyke to mark the frontier between England and Wales. It starts on the edge of the Severn bridge, and ends near Prestatyn. Since Offa's Dyke was built, however, a number of influences have caused English speech to penetrate into Wales. When the Normans built their chains of castles like those in Abergavenny and Cardiff, they reduced the influence of the Welsh in the lowland areas of central and South Wales. The highlands, on the other hand, remained Welsh, and the kingdom of Gwynedd in north-west Wales continued until it was conquered by King Edward I. The coming of industry to South Wales further weakened the Welsh language, especially in the upper valleys of east Glamorgan and Gwent which had been strongly Welsh.

The South Wales industrial area is by no means completely anglicized, however. Although the percentage of Welsh speaking is not high in Glamorgan, the actual numbers are considerable because the total population is large. Thus, over 200,000 people in Glamorgan are able to speak Welsh—nearly as many as in the strongly Welsh counties of north-west Wales.

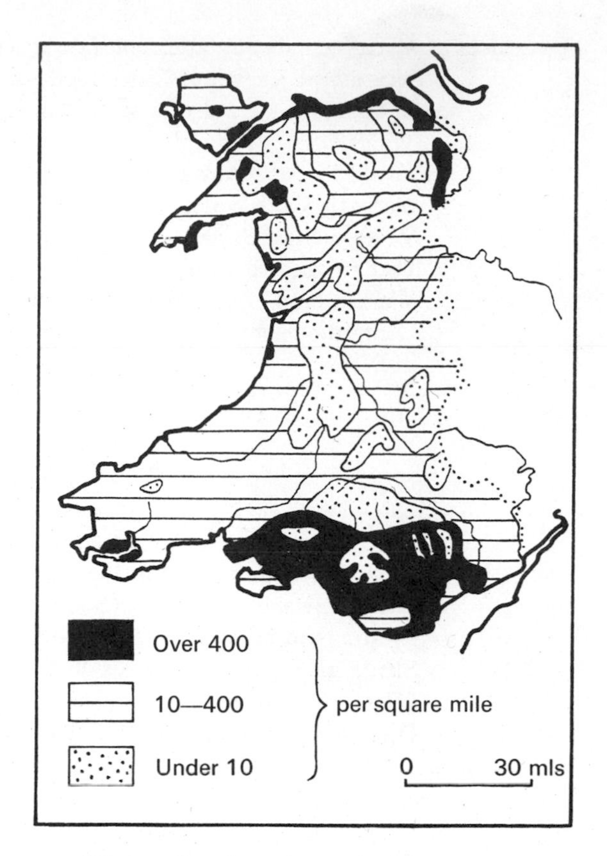

49. Wales: density of population

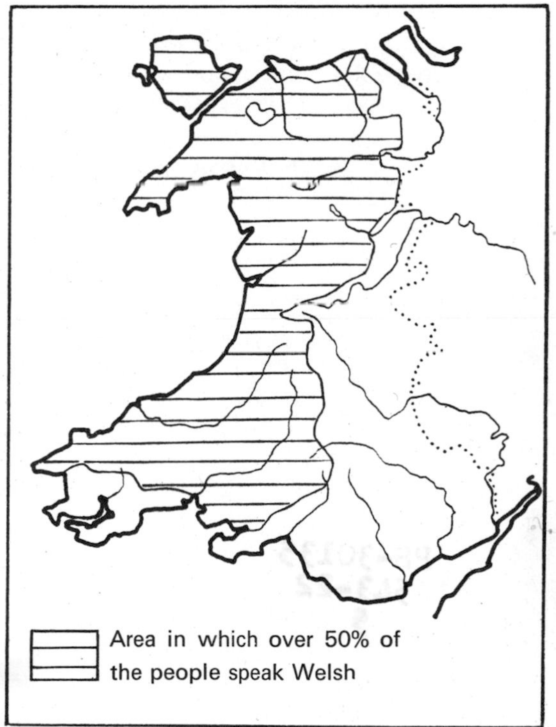

50. The Welsh-speaking population of Wales

Exercises 1. With the aid of an atlas find (a) the names of the rivers shown on the map (Fig. 39), (b) the location of Snowdon, Cader Idris, the Brecon Beacons and the Black Mountains.

2. Show, with the aid of a sketch map, the connection between rainfall and water supply in Wales.

3. Using Fig. 40, draw a series of columns proportional to the rainfall figures given for (a) four places in South Wales and (b) three places in North Wales arranged in order from east to west. Write a short explanation of what the diagrams show.

4. Using Fig. 42 as a guide, draw a sketch map showing the different kinds of farming in Wales. Write a brief explanation of what the map shows.

5. Using the figures given below, draw a simple diagram to represent the different uses of land in Wales:

Land use in Wales	*Areas in thousands of acres (1963).*
Forest	500
Rough grazing	1,600
Permanent grass	1,800
Arable land	860

Does the diagram confirm the impression given by the map (Fig. 42)?

6. Draw a sketch map of the South Wales coalfield showing: (a) coastline and main rivers, (b) limits of the coalfield, (c) Cardiff, Newport, Swansea and Llanelli.
Shade in the areas with most coal mines, distinguishing between anthracite and other coal. Why are there few mines in some parts of the coalfield?

7. Draw a sketch map of South Wales (excluding Pembrokeshire) showing: (a) coastline and main rivers, (b) limits of the coalfield, (c) chief towns and cities, (d) large steelworks, tinplate metal works, (e) trading estates and other factories, (f) main roads and railways.

8. Draw a sketch map of Milford Haven showing towns and oil refineries. Explain briefly why the oil refineries are located around Milford Haven.

9. How many different kinds of coastal towns could be seen in a journey around the coast from Flint to Newport? Give some examples.

10. Using Figs. 49 and 50, draw simple separate sketch maps to show each of the following, and write a short explanation of what each map shows: (a) areas where a large proportion of people speak Welsh, (b) densely populated and sparsely populated areas of Wales, and (c) the chief industrial areas of Wales.

PB-30135
543-22
5